1002 Wall Street
Los Angeles, CA 90015

Printed in the United States of America

2010
10 9 8 7 6 5 4 3 2 1

ISBN-13: 978-1-936206-10-0

Editorial
Dean Schaffer, Editor
Daniel Berdichevsky, Editor-in-Chief

WHAT IS A CRAM KIT?
A Word from the Editor

BETTER THAN THE TEXTBOOK YOU NEVER READ

GETTING STARTED

Everyone says that if you want to do well on an exam, cramming is not the way to do it. But not everyone has used a Cram Kit.

If you're reading this, you've already taken a big step toward improving your score. Cram Kits are DemiDec's signature study guide, focused on two types of information:

1. That which is most likely to be tested.
2. That which is most likely to be forgotten.

This Cram Kit is *not* a textbook. It does not aim to teach you everything you could ever want to know about a subject. Rather, Cram Kits give you a focused, approachable, and engaging path to success on the AP exam—especially if you're crunched for time.

So what are you waiting for? Get cramming!

THE CRAM FAMILY

The first generation of the Cram Family delivers Cram Kits and Cram Cards in the following subjects:

AP Psychology

AP Calculus

AP Chemistry

AP Biology

SAT I

AP U.S. History

AP English Language

AP Economics (Micro and Macro)

AP Spanish Language

CRAMMING FOR SUCCESS
A Word from the Author

BIOLOGY IN A NUTSHELL

THE BIG DAY

After a year of hard work, it's finally here! The AP Biology test may seem overwhelming, but this study guide will give you the tools you need to succeed. It covers every topic, big and small, and even the things you may have missed or forgotten.

It's a piece of cake! (Figuring out what to study, that is.) Why is it so easy? Check out the pie chart to the right.

- The biggest chunk of the AP Biology test deals with the structure and function of plants and animals
- Cell types, the cell cycle, and cell structure are key
- It's no surprise that evolution, heredity, and genetics are big players
- Don't forget ecology—sometimes it gets lost in the shuffle, but it's still a big part of the test

It's impossible to memorize everything. This book will help you prioritize topics and focus on concepts. Tables, graphics, and bolded terms will highlight what's important, and quizzes after each section will test your memory and your critical thinking.

Before the test, get a good night's sleep, eat a healthy breakfast, and put down your study aids. Let your brain process the information you've spent all this time studying. There's no need to worry. You're prepared!

PIECES OF THE AP PIE

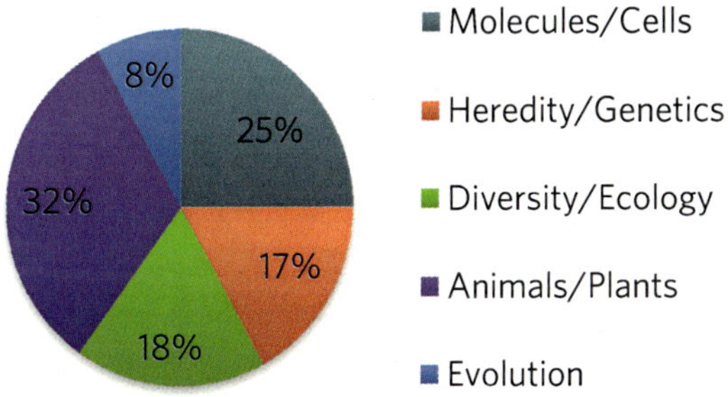

TIME IS TICKING!

- Biology is the study of life, so don't neglect yourself while you're studying!
- If you're pressed for time, focus on major steps in a process, not all the little details
- Learn the basic concepts (what is phosphorylation?) and then build from there (substrate-level or oxidative?)
- This is a college level course, so it's not easy!
- Keep moving, both during the test and while you study: don't get stuck on any one question or topic

CRAMMING FOR SUCCESS

How to Use This Book

A TERRIFIC TOOL

CRAMMING FOR SUCCESS

At their best, study aids clarify troublesome concepts or processes. (Can anyone say the Calvin cycle?) At their worst, they confuse.

This book strives to outline the AP Biology curriculum in a clear and engaging manner. You've already gotten a taste of the format. Information is organized by

- Bullets, numbers, and checks
- Graphics (with pretty colors, of course)
- Tables and charts
- Bolded terms and headings

At the top of each page, you will find two headings. The larger heading, in all capital letters, refers to an overarching section or theme of the curriculum—think of it as a chapter of the book (on this page, "CRAMMING FOR SUCCESS"). The smaller heading refers to the specific topic of that page (on this page, "How to Use This Book"). These two headings will help you organize material thematically in your brain.

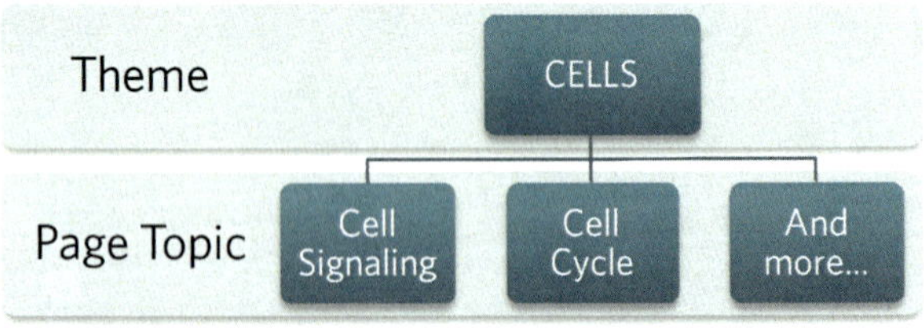

A quiz follows every page. It may cover material presented on the page previous or related information that wasn't included.

You'll find that the pages correspond to the percentages already mentioned—about a third of the pages, for example, deal with plant and animal structures.

CRUNCHING FOR SUCCESS

At the end of this book, you'll find a super-summary of the AP Biology course called the Crunch Kit. Use it as a quick reference, a last-minute refresher, or a high-speed review.

Following the Crunch Kit is the List of Lists, organized by topic. These lists introduce terms and their definitions that will probably be covered on the AP Biology exam. This section gives you a good chance to test your knowledge. Wait until you've reviewed the rest of the book and then see how many terms in the lists you can define without help. Most of the terms found in the lists are essential, so it's important to understand them fully before the test.

CRAMMING THE RIGHT WAY

KNOWING WHAT'S IMPORTANT

This book will help you with the major topics in the AP Biology course by reviewing terminology and different processes.

It's important to keep in mind that the concepts underlying the terminology are also important. The Calvin cycle, for example, is a complicated series of reactions. Memorizing each step of the cycle could make you crazy. Understanding that a plant takes carbon from the air and transfers it to an organic molecule is more important.

One could say the same thing for almost every topic in AP Biology. That doesn't mean you should ignore the details, though; just know where they come from.

THE MAJOR THEMES

In his textbook *Biology*, Neil Campbell introduces 11 themes that connect biology. Keep these concepts in mind while you're studying.

- **Cells** are the basic units of life
- Biological information is **inherited**
- Biological systems are organized in **hierarchies** that exhibit emergent properties
- Biological systems are **regulated** by feedback mechanisms
- Organisms **interact** with their environments
- **Energy** flows from the sun to organisms that need it to perform work
- Organisms are **diverse** but also **similar**
- **Evolution** is a core theme in biology, explaining how populations adapt to the environment and how new species form
- **Structure** and **function** are related at all levels of organization
- Science is furthered by **hypothesis-based inquiry**
- **Science**, **technology**, and **society** are connected

READY, GET SET, STUDY!

This is an exciting time in biology—the textbooks are changing quickly, and the field is growing and developing just as rapidly. But don't let that overwhelm you! The themes haven't changed. Life is still about order, adaptation, response, reproduction, energy, growth, and regulation. Relate what you're studying to these concepts. This will give you the good foundation you need to tackle any AP Biology test question. Good luck, and happy studying!

MOLECULES
Water

WATER, WATER EVERYWHERE

A SPECIAL KIND OF LIQUID

Water's chemical makeup gives the molecule its unique properties that sustain life on this planet. The oxygen atom in every water molecule has a slight negative charge while the hydrogen has a slightly positive charge (also called polarity), which causes hydrogen bonding with neighboring water molecules.

COHESION

Hydrogen bonding causes water molecules to cling to one another, creating tension that makes it difficult to break the surface of this liquid (surface tension). These bonds also cause adhesion, or the clinging of water to other substances.

TEMPERATURE MODERATION

Water's high specific heat (absorption of heat) and high heat of vaporization (conversion from liquid to gas) allows it stabilize temperatures.

1. Temperature increases
2. Hottest molecules evaporate
3. Cooler molecules left behind
4. Overheating prevented

INSULATION

One of water's strange properties is that it is less dense as a solid than as a liquid, which is why ice floats. Floating ice protects organisms living underwater from freezing.

SOLVENT

Many compounds dissolve in water, making it an excellent carrier of solutes (dissolved substances). Water breaks apart compounds and surrounds the exposed ions in hydration shells (water capsules), keeping them in solution.

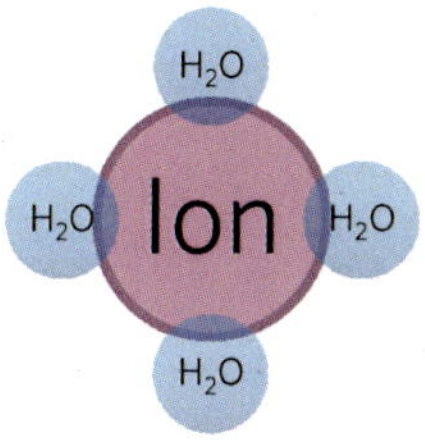

CHEMISTRY OF LIFE

SPLITTING UP

Water dissociates (separates) when a hydrogen ion (proton) breaks away from its parent water molecule and binds to a neighbor. The parent is now a hydroxide ion (OH-), and the neighbor becomes a hydronium ion (H_3O^+).

THE pH SCALE

Acids and bases upset hydrogen and hydroxide ions concentrations in a solution by adding or taking away hydrogen ions. Buffers minimize this change by receiving or donating hydrogen ions to a solution.

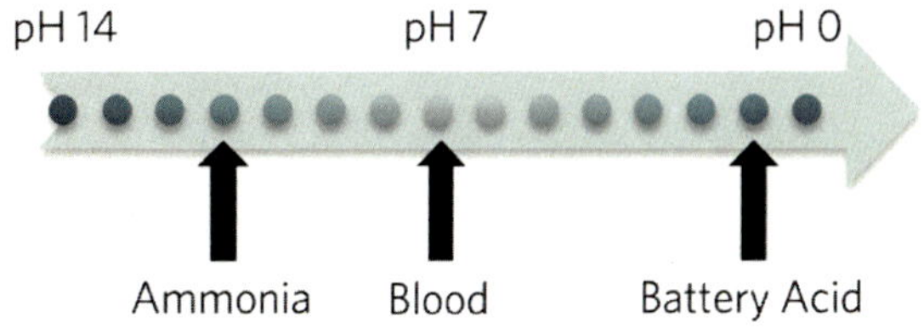

WATER AND CELLULAR PROCESSES

Water (and its component ions) plays a key role in crucial reactions that provide fuel for cells.

Electron transport chains	Oxygen is the last molecule in a chain to accept electrons, creating water.
Proton-motive force	Electron transport drives the pumping of H+ across membranes, creating a gradient.
Chemiosmosis	Gradient causes H+ to flow through ATP synthase, producing ATP.
Photosynthesis	Water donates two electrons to chlorophyll molecules to start photosynthesis.

CRAM QUIZ
Water

QUESTION 1

A solution with a pH of 9 has a hydroxide ion concentration of _______ and a hydrogen ion concentration of ______.

(A) 10^{-9}M, 10^{-14}M
(B) 10^{-10}M, 10^{-5}M
(C) 10^{-5}M, 10^{-14}M
(D) 10^{-5}M, 10^{-9}M
(E) 10^{-9}M, 10^{-5}M

QUESTION 2

H_2CO_3 is a

(A) weak base
(B) weak acid
(C) strong acid
(D) strong base
(E) neutral compound

QUESTION 3

Hydrophilic substances create colloids by

(A) dissolving in an aqueous solution
(B) being covalently bonded
(C) repelling large molecules
(D) weakly attracting water molecules
(E) crashing out of solution

QUESTION 4

Ice is less dense than most solids due to its

(A) covalent bonds
(B) crystalline structure
(C) bonding to three neighboring water molecules
(D) behavior at 4°C
(E) transient hydrogen bonds

QUESTION 5

Glycolysis

(A) uses four molecules of water
(B) produces four protons
(C) produces two molecules of water
(D) produces two molecules of ATP
(E) uses two protons

QUESTION 6

The proton motive force creates a difference in hydrogen ion concentration between the

(A) inside and outside of a cell
(B) compartments of the mitochondria
(C) cytosol and the mitochondria's inner membrane
(D) cristae and the cytosol
(E) mitochondrial matrix and the cytosol

QUESTION 7

Oxygen accepts the final electron in the electron transport chain that consists of how many protein complexes?

(A) one
(B) two
(C) three
(D) four
(E) five

QUESTION 8

The production of pyruvate kinase, the precursor to pyruvate, involves the removal of

(A) phosphate
(B) carbon
(C) hydrogen ions
(D) water
(E) oxygen

ANSWERS

1. D
2. B
3. D
4. B
5. C
6. B
7. D
8. C

MOLECULES
Carbon

THE BACKBONE OF LIFE

THE FOUNDATION OF DIVERSITY

Carbon compounds are found in all living organisms (hence the name *organic*). The diveristy of organic compounds comes from carbon atoms' structures: carbon contains six electrons total, with four valence electrons, each of which can pair with another electron.

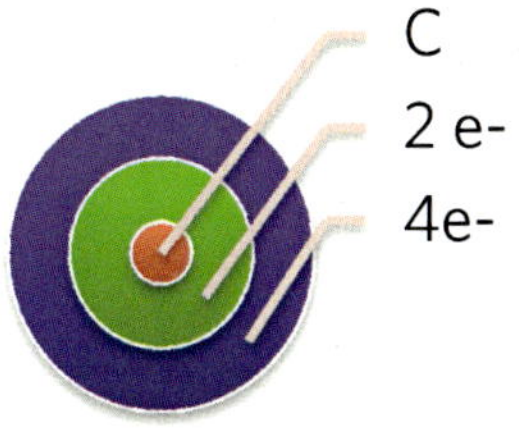

LUCKY NUMBER FOUR

Carbon's tetravalence (ability to bind with other atoms in four directions) makes it the perfect foundation for a diverse group of molecules. Chains of carbon and hydrogen atoms called hydrocarbons create a variety of structures.

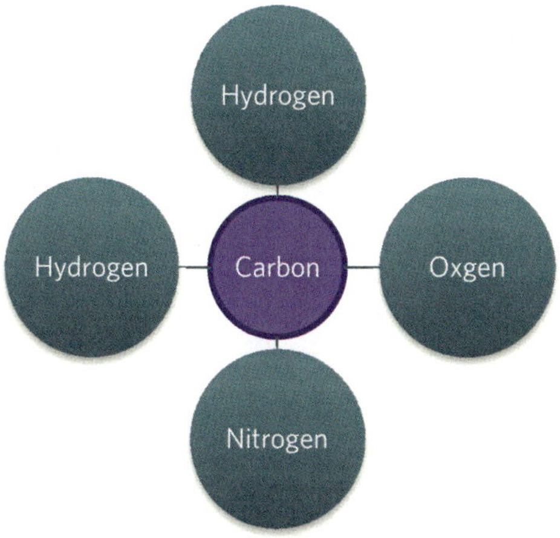

Note: Carbon often pairs with nitrogen, oxygen, and hydrogen based on the rules of covalent bonding.

SPOTLIGHT ON CHEMICAL BONDING

The complexity of life centers around the ability of atoms to bind together and form diverse molecules. Some molecular interactions are strong and others weak, depending on their electronegativities (electron attractions).

Strong Bonds	Weak Bonds
Covalent	Hydrogen
Ionic	Van der Waals

IT'S ALIVE

VERY VERSATILE

The functional groups attached to carbon compounds usually take part in reactions and contribute to the versatility and functionality of organic molecules.

POLYMERS AND THEIR FUNCTIONAL GROUPS

The body contains many different kinds of large molecules called polymers that perform different functions in cells. *(Note: fats are not polymers.)*

Functional Group	Polymer	Job	Example
Carbonyl	Saccharide	Fuel	Glucose
Hydroxyl	Saccharide	Storage Structure	Glycogen Cellulose
Hydroxyl	Fats	Energy storage	Triacylgly -cerol
Phosphate	Phospho-lipid	Membrane compo-nent	Phospha-tidyl-choline
Varies	Steroid	Molecular Precursor	Cholest-erol
Carboxyl Amino	Amino acid	Building blocks of proteins	Glycine
Phosphate	Nucleic acid	Heredity	DNA RNA
Sulfhydryl	Protein	Tertiary structure	Collagen

POLYMER BONDING FUELED BY DEHYDRATION

- **Glycosidic linkage**: the removal of water (dehydration) forms a covalent bond between two monosaccharides
- **Ester linkage**: dehydration between a glycerol and a fatty acid forms fat
- **Peptide bonds**: dehydration links carboxyl and amino groups of peptides, forming polypeptide chains

CRAM QUIZ
Carbon

QUESTION 1

Chitin is a structural polysaccharide found in the exoskeletons of insects. It contains which of the following functional groups?

(A) carboxyl and carbonyl
(B) carbonyl and hydroxyl
(C) carboxyl and hydroxyl
(D) sylhydryl and carboxyl
(E) sulhydryl and carbonyl

QUESTION 2

Organic compounds exhibit branching due to carbon atoms'

(A) number of filled electron shells
(B) electronegativity
(C) valence
(D) functional groups
(E) covalent bonding

QUESTION 3

Organic phosphates such as glycerol phosphate have which of the following properties?

(A) make molecules positively charged
(B) form a basic solution
(C) singly bond phosphate to four oxygen molecules
(D) catalyze reactions
(E) ionize phosphate groups

QUESTION 4

Chemical equilibrium of different forms of glucose prefers

(A) linear structures
(B) branched structures
(C) alpha structures
(D) beta structures
(E) ring structures

QUESTION 5

Name the location of the carbonyl group in carbon skeletons of ketones and aldehydes and its chemical composition, respectively.

(A) end of skeleton, inside skeleton, -COOH
(B) end of skeleton, middle of skeleton, -CO
(C) inside skeleton, end of skeleton, -COOH
(D) inside skeleton, end of skeleton, -CO
(E) inside skeleton, middle skeleton, -CO

QUESTION 6

Isomers have the same number of atoms of the same elements in different arrangements. Geometric isomers

(A) are mirror images of each other
(B) have same covalent bonds arranged differently
(C) come in *cis* and *trans* forms with different covalent bonds
(D) have the arrangement of different forms of pentane
(E) have asymmetric carbons

QUESTION 7

A saturated fatty acid is solid at room temperature. Under the same conditions, an unsaturated fatty acid is not solid, due to

(A) *trans* double bonding
(B) tight packing of molecules
(C) kinks in the tails
(D) addition of hydrogen atoms
(E) a natural, not synthetic, process

QUESTION 8

A protein's tertiary structure involves

(A) interactions of polypeptide backbone
(B) hydrophilic interactions
(C) hydrogen bonding
(D) peptide bonding
(E) hydrophilic interactions

ANSWERS

1. B
2. C
3. E
4. E
5. D
6. B
7. C
8. B

MOLECULES
Metabolism

SPEED DEMON

CRAZY CATALYSTS

Enzymes are proteins that catalyze (speed up) reactions by lowering the reaction's activation energy without being used up in the process.

Enzymes Change Reaction Rate

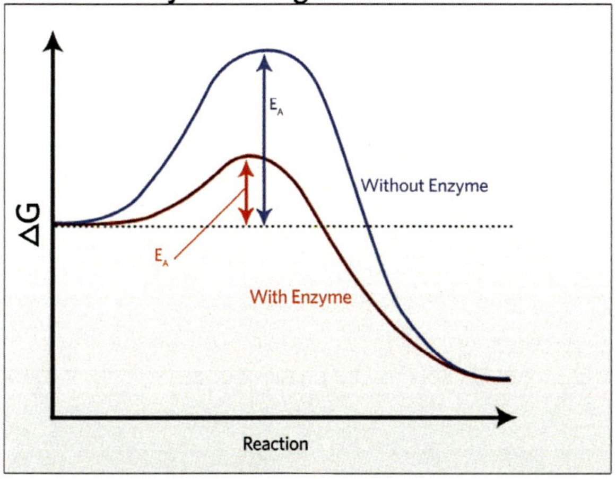

HOW'D YOU DO THAT?

An enzyme speeds up an endergonic reaction by making it exergonic. It lowers a reaction's activation energy by

1. Changing the substrate's orientation
2. Stretching the substrate's bonds
3. Creating a favorable environment (for example, low pH)
4. Participating in the reaction

KEY TERMS

Activation energy	Energy needed to start a reaction
Exergonic reaction	Releases free energy; ΔG negative
Substrate	Reactant that binds temporarily to enzyme
Active site	Substrate binding site on enzyme

FITS LIKE A GLOVE

An enzyme's active site exposes a few functional groups to a substrate. When binding occurs, the active site changes shape to fit the substrate (induced fit) and is held in place by weak bonds. The substrate converts to a product and releases the enzyme, which is now available for another reaction.

CONTROL MECHANISMS

LOCAL CONDITIONS

Factors that affect enzymatic activity help cells regulate the reactions that make up an organism's metabolism.

COFACTORS

Cofactors are non-proteins that bind permanently or temporarily to an enzyme and affect substrate binding.

INHIBITORS

Inhibitors prevent an enzyme from catalyzing a reaction by binding to the active site (competitive inhibition) or binding to another site that changes the active site's conformation (noncompetitive inhibition).

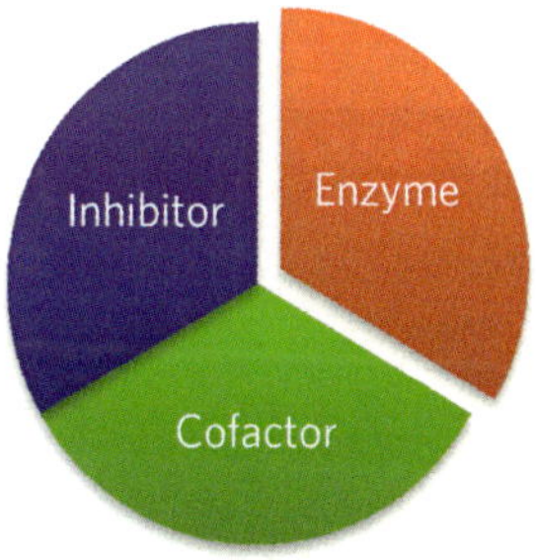

ALLOSTERIC REGULATION

Allosteric regulators bind to an enzyme at a location other than the active site and stabilize its conformation. Activators stabilize a working enzyme; inhibitors stabilize an inactivated enzyme.

FEEDBACK INHIBITION

Feedback inhibition regulates metabolism by having a reaction's product inhibit the enzyme that catalyzes it.

CHANGES TO PROTEIN FOLDING

Proteins have intricate structures that allow them to perform specialized functions like catalyzing reactions. Temperature and pH changes disrupt this structure and cause the protein to denature.

Primary structure	Amino acid sequence
Secondary structure	Hydrogen bonding forms beta sheet and alpha helices
Tertiary structure	Side chain interactions create functional conformation
Quaternary structure	Overall structure

CRAM QUIZ
Metabolism

QUESTION 1

The change in free energy (ΔG) of a system is calculated by which equation, where T equals absolute temperature, H equals enthalpy, and S equals entropy?

(A) $\Delta G = S\Delta T - H$
(B) $\Delta G = \Delta H + TS$
(C) $\Delta G = \Delta S - \Delta HT$
(D) $\Delta G = \Delta H - T\Delta S$
(E) $\Delta G = \Delta T - H\Delta S$

QUESTION 2

Together, the first and second laws of thermodynamics explain that

(A) energy can be transferred out of a system as the system becomes more ordered
(B) energy is created by increasing the randomness of a system
(C) energy transfer increases the disorder of a system
(D) chaos is inherent in every system and is not dependent on energy
(E) energy cannot be created or destroyed but causes disorderliness

QUESTION 3

Cellular respiration is a catabolic pathway because it breaks down glucose. An example of an anabolic pathway is

(A) the production of polypeptides from amino acids
(B) the production of glucose from glycogen
(C) a series of hydrolytic reactions
(D) the production of triacylglycerol
(E) the production of nucleosides from nucleotides

QUESTION 4

A chaperonin from *E. coli* contains two proteins that stabilize the conformations of other proteins by

(A) forming a protective shield around the protein that gives it time to renature
(B) binding to the protein and changing its conformation
(C) sealing the protein in a hydrophilic environment
(D) repairing its hydrogen bonding
(E) catalyzing the reaction that causes the protein to refold

QUESTION 5

Free energy is related to a system's stability and ability to perform work because a

(A) more stable system has less free energy and less ability to perform work
(B) less stable system has less free energy and less ability to perform work
(C) more stable system has more free energy and less ability to perform work
(D) less stable system has more free energy and more ability to perform work
(E) more stable system has less free energy and more ability to perform work

QUESTION 6

Energy coupling uses an exergonic reaction to fuel an endergonic one. ATP is the primary intermediary in this process because it

(A) lowers the activation energy of other reactions
(B) contains phosphate groups that bind to other molecules
(C) undergoes dehydration readily and fuels other reactions
(D) lowers the free energy of a system via anabolism
(E) has a high chemical energy

QUESTION 7

The cell performs three types of work: mechanical, transport, and chemical. ATP fuels these processes by doing what to reactants?

(A) catalyzing them
(B) hydrolyzing them
(C) catabolyzing them
(D) phosphorylating them
(E) polymerizing them

ANSWERS

1. D
2. C
3. A
4. C
5. B
6. B
7. D

CELLS
Major Cell Types

BUILDING BLOCKS

SOLID FOUNDATION

Cells are the structural and functional units of living organisms and fall into one of two categories: prokaryotic and eukaryotic.

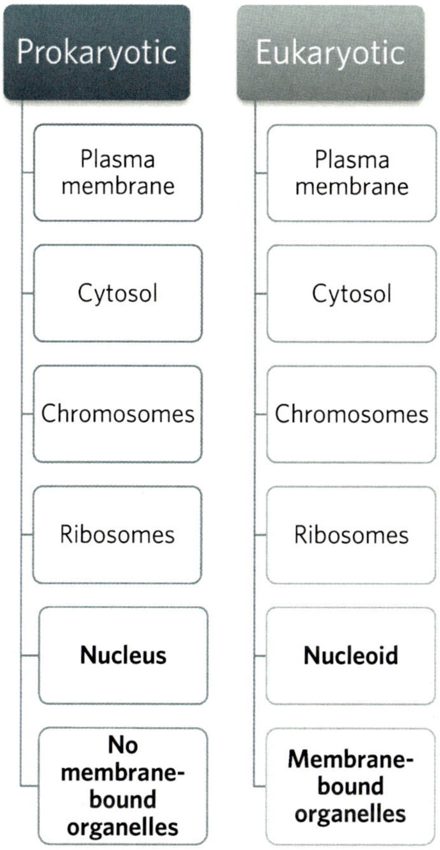

KEY FEATURES

Prokaryotic Cells	*Eukaryotic Cells*
Generally smaller	Usually bigger
Found in bacteria and archaea	Found in plants, animals, protists, and fungi
DNA is not separated from the rest of the cell	Organelles with membranes, like mitochondria, chloroplasts, and nuclei

GREAT BARRIERS

PLASMA MEMBRANE

The plasma membrane is a barrier that separates a cell from the outside environment and has many characteristics:

- ✓ Selectively permeable
- ✓ Transports nutrients into the cell
- ✓ Disposes of waste
- ✓ Double layered (also called a bilayer)
- ✓ Hydrophobic ("water-hating") interior
- ✓ Hydrophilic ("water-loving") exterior

ENDOMEMBRANE SYSTEM

A maze of membranes called the endomembrane system divides the inside of a eukaryotic cell into compartments. The cell carries out different functions in each compartment.

Major Functions	Protein synthesis and transport, detoxification, lipid metabolism and transport
Components	Smooth ER, rough ER, Golgi apparatus, lysosomes, vacuoles, ribosomes

SHARED ANCESTRY

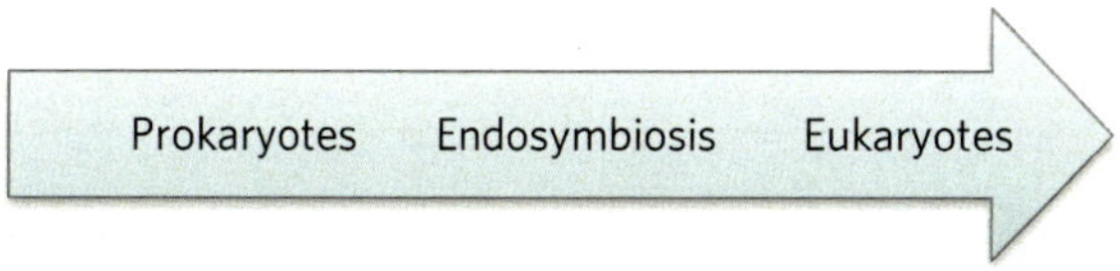

Scientists think that eukaryotes evolved from prokaryotes about 2 to 3 million years ago by endosymbiosis. By this strange process, prokaryotes engulfed other prokaryotes that became eukaryotic organelles (mitochondria and chloroplasts).

The first multicellular organisms were colonies of cells that self-replicated. Some cells had specialized functions like nitrogen fixation. More complex eukaryotic cells appeared about 1.7 billion years ago in the form of simple algae.

CRAM QUIZ
Major Cell Types

QUESTION 1

The nuclear lamina lines the inside of the nuclear envelope with

(A) protein filaments
(B) nuclear matrix
(C) microtubules
(D) pores
(E) large macromolecules

QUESTION 2

The nucleolus functions in

(A) messenger RNA production
(B) DNA replication
(C) membrane synthesis
(D) ribosomal synthesis
(E) DNA condensation

QUESTION 3

The smooth ER stores what ions in its lumen (cavity)?

(A) potassium
(B) hydrogen
(C) calcium
(D) chloride
(E) sodium

QUESTION 4

The *trans* face of the Golgi apparatus

(A) is closest to the ER
(B) ships molecules to the ER
(C) receives molecules from the ER
(D) pinches off vesicles
(E) modifies proteins and other molecules

QUESTION 5

Ribosomes are made of

(A) endoplasmic reticulum
(B) carbohydrates and lipids
(C) DNA and proteins
(D) lipids and carbohydrates
(E) RNA and proteins

QUESTION 6

The cells walls of prokaryotes

(A) contain a jelly like substance
(B) protect them in a hypertonic environment
(C) prevent plasmolyzing
(D) prevent bursting
(E) are made of cellulose

QUESTION 7

Which cellular structure is NOT convoluted to increase cell surface area?

(A) rough ER
(B) nucleus
(C) smooth ER
(D) Golgi apparatus
(E) chloroplast

QUESTION 8

The ancestral precursors to mitochondria were

(A) aerobic heterotrophic prokaryotes
(B) aerobic autotrophic prokaryotes
(C) plastids
(D) aerobic eukaryotes
(E) anaerobic prokaryotes

ANSWERS

1. A
2. D
3. C
4. D
5. E
6. D
7. B
8. A

CELLS
Eukaryotic Cell Components

ANIMAL OR VEGETABLE?

DIFFERENT "LITTLE ORGANS"

Plant and animal cells contain different organelles that reflect their diverse functions. Central vacuoles and cell walls, for example, create plant cells' rigid structure.

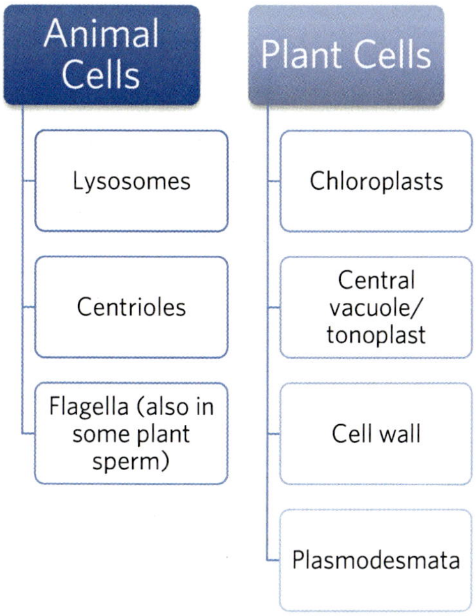

A CLOSER LOOK

Organelle	*Function*
Lysosome	Digestive sac that contains hydrolytic enzymes
Centrioles	Nine sets of three microtubules formed in a ring and located inside the centrosome
Central vacuole	Holds fluid called cell sap that contains nutrients or waste
Tonoplast	Selective barrier surrounding central vacuole
Cell wall	Made of cellulose and proteins that surround the plasma membrane

ENERGETIC ORGANELLES

MITOCHONDRIA

Cellular respiration produces ATP by breaking down fuel molecules (carbohydrates and lipids); it takes place in the mitochondria. The mitochondria's double membrane separates the reaction into several steps.

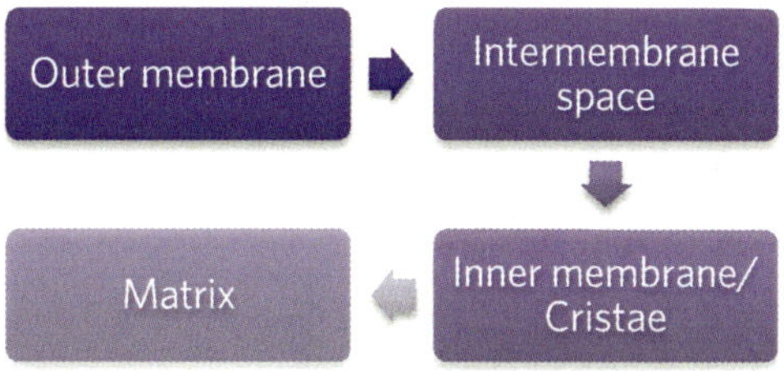

- The outer membrane actively transports pyruvate into intermembrane space
- The citric acid cycle and oxidative phosphylation take place within the mitochondria
- The outer membrane transports ATP into the cytosol

CHLOROPLASTS

In a series of reactions, photosynthesis converts light energy and carbon dioxide into glucose and oxygen; it takes place in the chloroplasts.

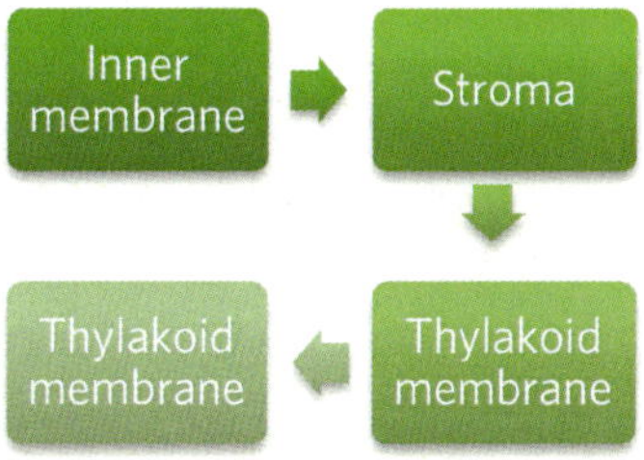

- Light reactions take place in the thylakoid membranes
- Calvin cycle occurs in the fluid-filled stroma

PEROXISOMES

Single membrane organelles called peroxisomes break down substrates by transferring hydrogen to oxygen and function in...

- Fatty acid metabolism
- Alcohol and poison detoxification
- Hydrogen peroxide production and breakdown

CRAM QUIZ
Eukaryotic Cell Components

QUESTION 1

Plants and animal cells have what in common?

(A) mitotic division
(B) maintenance of structure
(C) energy source for ATP production
(D) organelles
(E) flagella

QUESTION 2

The function of the centrioles is

(A) to pull apart sister chromatids during mitosis
(B) to organize the cytoskeleton
(C) to produce microtubules
(D) unknown
(E) to anchor membrane proteins

QUESTION 3

Free floating ribosomes in the mitochondrial matrix are found with

(A) chromosomes
(B) chromatin
(C) circular DNA
(D) RNA
(E) chromatids

QUESTION 4

The infoldings called cristae are found in the

(A) thylakoid membrane
(B) mitochondrial outer membrane
(C) Golgi apparatus
(D) granna
(E) mitochondrial inner membrane

QUESTION 5

The central vacuole in plant cells

(A) is surrounded by a tonoplast
(B) is another name for a tonoplast
(C) performs the same function as the tonoplast
(D) contains water, as does the tonoplast
(E) is in younger plants

QUESTION 6

Plastids have many functions, including all of the following EXCEPT

(A) carrying out cellular respiration
(B) storage of starch
(C) glucose production
(D) giving fruit their color
(E) performing chemiosmosis

QUESTION 7

Peroxisomes are produced by the

(A) ER
(B) Golgi apparatus
(C) plasma membrane
(D) free proteins
(E) nuclear envelope

QUESTION 8

Chloroplasts are found in

(A) plants and prokaryotes
(B) all plants and protists
(C) all plants
(D) plants and algae
(E) plants and some protists

ANSWERS

1. E
2. D
3. C
4. E
5. E
6. A
7. D
8. D

CELLS
Cytoskeleton and Movement

THE TINIEST SKELETONS

DYNAMIC STRENGTH

The cell's skeleton (cytoskeleton) gives the cell its shape, helps in movement, and provides structural support by constantly changing the web of fibers crisscrossing the cell.

FLEXIBLE FIBERS

The cytoskeleton is made up of three different fibers that perform various functions.

Fiber	*Structure*
Microtubule	Tubulin dimmers circling a hollow tube
Microfilament	Two intertwined actin polymers
Intermediate filaments	Keratin proteins supercoiled into cables

Eukaryotic Cytoskeleton (actin, microtubules, nuclei)

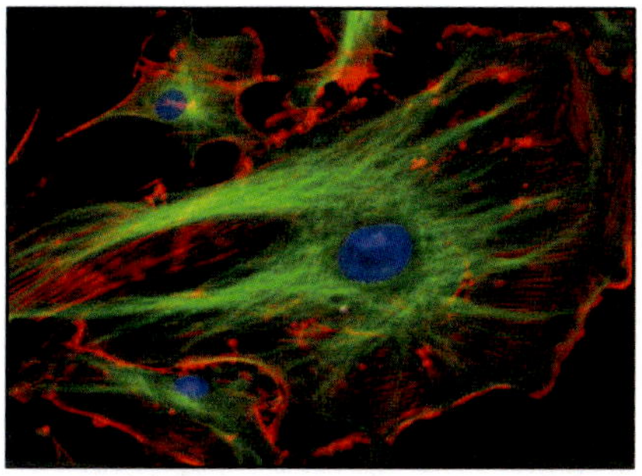

SPOTLIGHT ON FUNCTION

Microtubules
- Cell shape
  - Maintain
- Movement
  - Chromosomes
  - Cells
  - Organelles

Microfilaments
- Cell shape
  - Change
  - Maintain
- Movement
  - Pseudopodia
  - Cytoplasmic streaming
- Other Functions
  - Muscle contract
  - Form cleavage furrow

Intermediate fibers
- Cell Shape
  - Maintain
- Other Functions
  - Anchor
  - Form nuclear lamina

BODIES IN MOTION

TREMENDOUS TUBES

Microtubules crisscross the eukaryotic cell, grow out from the centrosome, and separate chromosomes during division.

SAME STRUCTURE, DIFFERENT MOVES

Microtubules make up eukaryotic **cilia** and **flagella**, cell structures responsible for movement. Although their motions differ, cilia and flagella have similar structures.

- A ring of nine microtubule doublets circles a pair of microtubules (**9 + 2 pattern**)

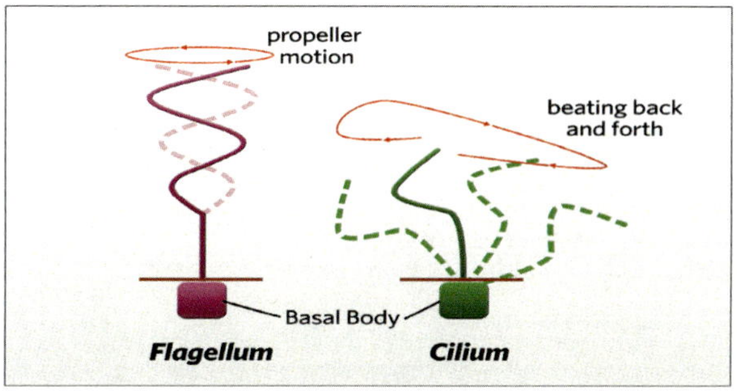

ONE STEP AT A TIME

ATP powers cilia and flagella movement as a motor protein called **dynein** "walks" up the microtubules, causing them to bend.

PERFECT PAIR: ACTIN AND MYOSIN

DYNAMIC DUO

Actin filaments interact with other proteins called myosin to create movement in amoebas, plants, and humans.

Muscle Contraction
- Actin attached to parrallel myosin filaments
- Myosin pulls actin
- Muscle cells shorten

Amoeboid Movement
- Myosin and actin cause contraction at one end
- Actin filaments at other end break down
- Pseudopodia extends

Cytoplasmic Streaming
- Actin and myosin interact
- Cytoplasm moves around plant cell

CRAM QUIZ
Cytoskeleton and Movement

QUESTION 1

The interaction between motor proteins and the cytoskeleton are NOT responsible for what type of movement?

(A) neurotransmitter vesicles migrating to the axon hillock
(B) contraction of muscle cells
(C) movement of vesicles from the ER to the Golgi apparatus
(D) nuclear rearrangement in nondividing cells
(E) prokaryotic movement toward a stimuli

QUESTION 2

When an external force compresses a cell, which type of filament maintains cell shape?

(A) microtubule
(B) microfilament
(C) intermediate filament
(D) myosin filament
(E) flagellin filament

QUESTION 3

Vesicles travel from the endoplasmic reticulum to the *cis* face of the Golgi apparatus. At the *trans* face of the Golgi, other vesicles are guided by what filaments to the plasma membrane?

(A) microtubules
(B) microfilaments
(C) intermediate filaments
(D) myosin filaments
(E) flagellin filaments

QUESTION 4

The arrangement of microtubules in the outer ring of centrioles, basal bodies (of cilia or flagella), and flagella are, respectively,

(A) nine sets of two microtubules, nine triplets, and nine doublets
(B) nine doublets, nine sets of three microtubules, and nine doublets
(C) nine sets of three microtubules, nine doublets, and nine doublets.
(D) nine triplets, nine sets of three microtubules, and nine doublets
(E) nine triplets, nine doublets, and nine doublets

QUESTION 5

Experiments that destroy tubulin and actin, but leave keratin fibers suggest that

(A) microfilaments are more dynamic fibers
(B) intermediate filaments fix organelles in a particular location
(C) skin cells are made up of mostly microtubules
(D) intermediate filaments rearrange quickly
(E) intermediate filaments are responsible for cell shape

QUESTION 6

In amoeboid movement, actin filaments opposite the trailing (moving) end of the cell

(A) break down into subunits
(B) remain filaments
(C) interact with myosin
(D) form a network
(E) are found in the sol

QUESTION 7

Branching filaments directly inside the cell's plasma membrane are

(A) microtubules
(B) microfilaments
(C) intermediate filaments
(D) myosin filaments
(E) flagellin filaments

QUESTION 8

Dynein walking does not cause microtubules to slide past each other because

(A) dimers are attached to each other
(B) dynein arms keep them together
(C) they are restrained by proteins
(D) they are held together by radial spokes
(E) they are bound by microfilaments

ANSWERS

1. E
2. A
3. A
4. C
5. B
6. D
7. B
8. C

CELLS
Cellular Connections

KEEPING IN TOUCH

TIGHT KNIT

Cells form close-knit networks with the help of proteins that penetrate (integral) or are attached to (peripheral) the plasma membrane.

BINDING PROTEINS (INTERCELLULAR JUNCTIONS)

- Perforated channels called **plasmodesmata** connect the cytosol of adjacent *plant cells* and create a continuous environment; proteins embedded in *animal cell* membranes form pores called gap junctions through which small molecules pass
- Proteins bind membranes of adjacent *animal cells* (usually epithelial), creating a **tight junction** that prevents leakage
- **Desmosomes** anchor *animal cell* membranes together with intermediate keratin filaments

OTHER MEMBRANE PROTEINS

Function	*Details*
Facilitated diffusion	Proteins help substances passively diffuse through a membrane by carrying them across or creating a channel. ▪ Aquaporins (water transport) ▪ Ion channels ▪ Gated channels (change conformation to facilitate transport)
Active transport	Cell expends energy to transport a substance through a membrane. ▪ Sodium-potassium pump (3 Na^+ ions out; 2 K^+ ions in) ▪ Electrogenic pump (creates voltage) ▪ Proton pump (protons out)
Metabolism	Enzymes catalyze reactions.
Communica-tion	Chemical messenger changes receptor (protein) conformation.
Recognition	Glycoproteins establish a cell's identity.
Structure	Proteins adhere to cytoskeleton to maintain the cell's shape.

THE SPACE BETWEEN

BEYOND THE PLASMA MEMBRANE

Structures outside the plasma membrane of plant and animal cells maintain structure and help in communication.

CELL WALLS

Location	Plant cells
Major Functions	Protection, structural integrity; prevent lysis
Composition	Matrix of cellulose microfibrils, proteins, and polysaccharides
Components	Primary cell wall, middle lamella, secondary cell wall

EXTRACELLULAR MATRIX (ECM)

Location	Animal cells
Major Functions	Regulation, communication, structural integrity
Composition	Glycoproteins such as collagen, proteoglycans, fibronectin, integrins

POSITIVELY PROKARYOTIC

SPOTLIGHT ON GRAM STAINING

Prokaryotes fall into two different categories based on cell wall composition.

Gram positive
- More peptidoglycan
- Dyes deep violet
- Simpler structure

Gram negative
- Less peptidoglycan
- Dyes pink or red
- Structurally complex
- Contains outer membrane with toxic lipopolysaccharides

CRAM QUIZ
Cellular Connections

QUESTION 1

In the plasma membranes of animal cells, cholesterol

(A) increases fluidity in moderate temperatures
(B) decreases fluidity at all temperatures
(C) creates pockets of space at low temperatures
(D) reduces phospholipid movement
(E) straightens phospholipids' fatty acid tails

QUESTION 2

When a plasma membrane is frozen and fractured into two layers, it is possible to see which

(A) side is hydrophobic
(B) layer contains cholesterol
(C) layer is more densely packed
(D) layer contains proteins
(E) layer contains carbohydrates

QUESTION 3

Which molecule cannot easily diffuse through a plasma membrane?

(A) oxygen
(B) hydrocarbon
(C) glucose
(D) water
(E) carbon dioxide

QUESTION 4

In a hypertonic environment, a cell will

(A) burst
(B) lose solutes
(C) become hypotonic
(D) lose water
(E) osmoregulate

QUESTION 5

Proton pumps create a voltage across a membrane by transporting

(A) water into the cell
(B) calcium ions out of the cell
(C) hydrogen ions out of the cell
(D) calcium into the cell
(E) hydrogen ions into the cell

QUESTION 6

The process by which a cell consumes extracellular fluid in vesicles is called

(A) phagocytosis
(B) pinocytosis
(C) exocytosis
(D) receptor-mediated endocytosis
(E) diffusion

QUESTION 7

Cellular junctions that allow animal cells to communicate are called

(A) tight junctions
(B) pili
(C) desmosomes
(D) plasmodesmata
(E) gap junctions

QUESTION 8

In the ECM of animal cells, microfilaments are

(A) within the plasma membrane
(B) attached to the external side of the membrane
(C) attached to the inside of the membrane
(D) with collagen fibers
(E) in the proteoglycan complex

ANSWERS

1. C
2. D
3. C
4. D
5. C
6. B
7. C
8. C

CELLS
Photosynthesis

LET THERE BE LIGHT

SOLAR POWERED

Light reactions put the *photo* in photosynthesis. Photosystems in the thylakoid membrane use light energy to drive NADPH and ATP synthesis.

PHOTOSYSTEM BREAK DOWN

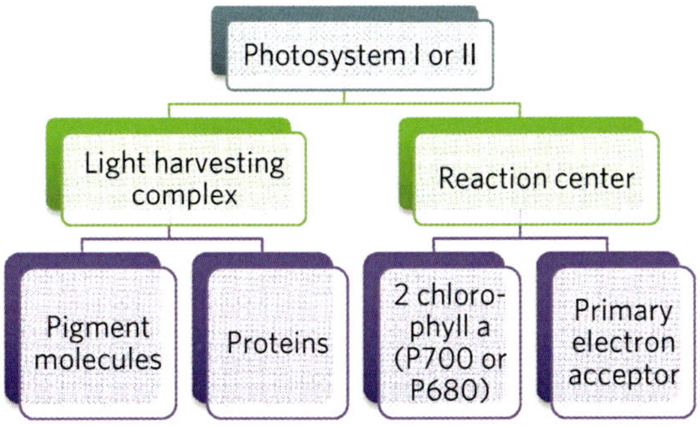

Note: Photosystem I contains the chlorophyll a molecule P700 (which absorbs light at the wavelength 700 nm). PSII contains P680 chlorophyll a.

CYCLICAL OR NOT?

After light energizes photosystems, electrons flow from the primary electron acceptor to an electron transport chain. Electron flow is cyclic or noncyclic, depending on which photosystem is excited.

CYCLIC ELECTRON FLOW IN PSI

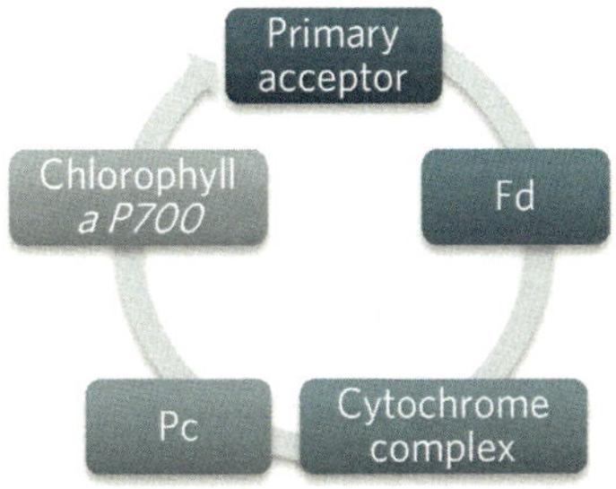

Noncyclic Electron Flow from PSII to PSI

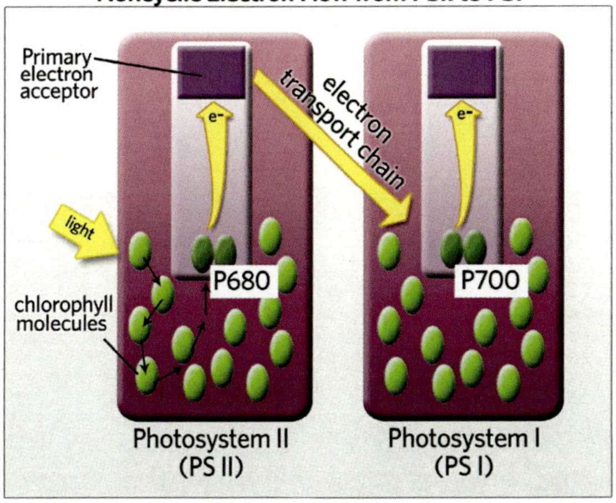

GETTING YOUR SUGAR FIX

SWEET PROCESS

ATP and NADPH from the light reactions travel to the Calvin cycle, the second stage of photosynthesis. The Calvin cycle fixes carbon to produce the sugar molecules the plant uses for energy.

Note: Carbon fixation takes inorganic carbon (from CO_2 in this case) and transfers it to an organic molecule.

THE CALVIN CYCLE IN THREE PHASES

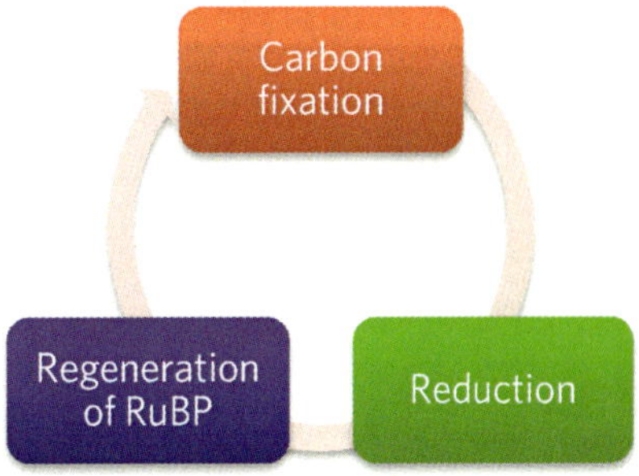

A CLOSER LOOK

- Carbon fixation—3 CO_2 are incorporated into 3 RuBP by the enzyme rubisco, creating an unstable molecule that splits in two
- Reduction—ATP donates phosphates to each molecule; NADPH donates electrons, creating sugar molecules (one is released to the cell)
- Regeneration—5 sugar molecules rearranged to form 3 RuBP with addition of phosphates from ATP

CLEVER ADAPTATIONS

C4 and CAM plants adapt to dry conditions by separating carbon fixation and the Calvin cycle spatially or temporally.

C4 versus CAM plants

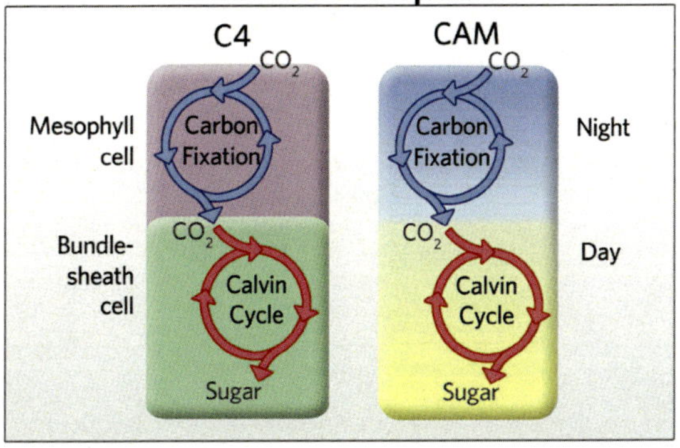

CRAM QUIZ
Photosynthesis

QUESTION 1

In photosynthesis, NADPH serves as a source of

(A) phosphate
(B) hydrogen ions
(C) electrons
(D) chemical energy
(E) sugar

QUESTION 2

The oxygen given off by plants during photosynthesis originates from

(A) glucose
(B) proteins
(C) carbon dioxide
(D) water
(E) Calvin cycle

QUESTION 3

What two molecules cycle from the light reactions to the Calvin cycle?

(A) oxygen and water
(B) oxygen and ATP
(C) water and ATP
(D) ATP and NADPH
(E) ADP and NADP+

QUESTION 4

Chlorophyll *a* and *b* do NOT differ in the

(A) wavelengths of light they absorb
(B) functional groups attached to their polyphyrin rings
(C) location of their reaction centers
(D) colors
(E) absorption spectra

QUESTION 5

Chloroplasts and mitochondria make ATP from chemiosmosis. How do these reactions differ in the two structures?

(A) They use different molecules to extract electrons.
(B) Chloroplasts do not use proton-motive force.
(C) Hydrogen ions are pumped into the mitochondrial matrix versus the stroma.
(D) Different forms of energy drive the reactions.
(E) ATP synthases are structurally dissimilar.

QUESTION 6

The sugar produced by the Calvin cycle is

(A) glucose
(B) glyceraldehyde 3-phosphate
(C) 1,3 bisphosphoglycerate
(D) aldehyde-3-phosphate
(E) glyceraldehyde

QUESTION 7

Rubisco is an inefficient enzyme that fixes oxygen as readily as carbon dioxide. This reaction is called

(A) photorespiration
(B) photophosphorylation
(C) crassulacean acid metabolism
(D) C4 metabolism
(E) oxidative respiration

QUESTION 8

C4 plants are so named because the first product in the pathway is

(A) PEP
(B) Malate
(C) Oxaloacetate
(D) Glyceraldehyde
(E) RuBP

ANSWERS

1. C
2. D
3. D
4. C
5. D
6. B
7. A
8. C

CELLS
Cell Signaling

3, 2, 1, CONTACT!

CAN YOU HEAR ME NOW?

Cells communicate through signal transduction pathways—a series of reactions that receive, transmit, and respond to chemical signals.

STEP 1: RECEPTION

A chemical messenger (also called a ligand) binds to a receptor on the cell surface that complements its shape, changing its conformation. This activates the receptor and causes it to interact with other molecules.

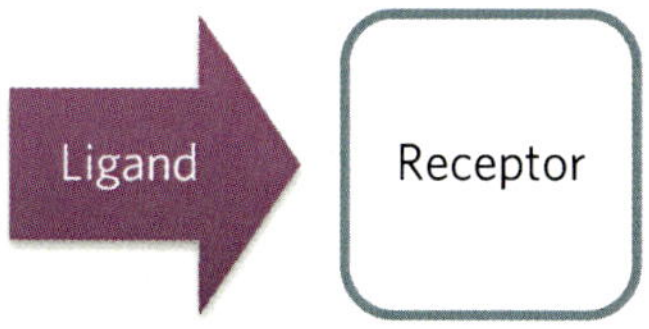

STEP 2: TRANSDUCTION

An activated receptor starts a cascade of reactions, wherein a protein kinase usually phosphorylates (donates a phosphate group from ATP) to another kinase, and so on.

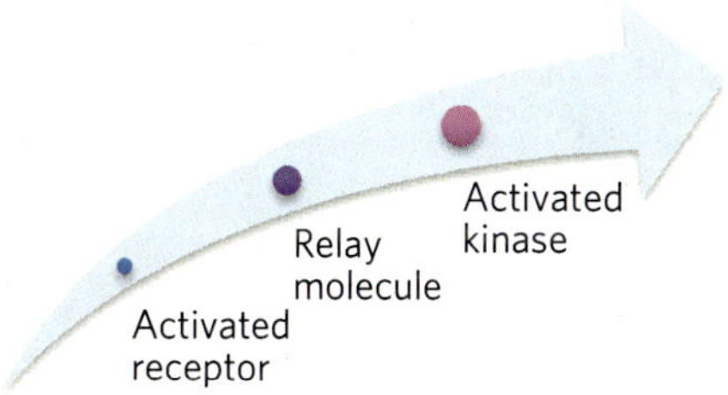

STEP 3: RESPONSE

Signal molecules cause changes in transcription or cytoplasmic activities. An ion channel in the plasma membrane may open, close, or stimulate the synthesis of a particular enzyme that will affect cell functions.

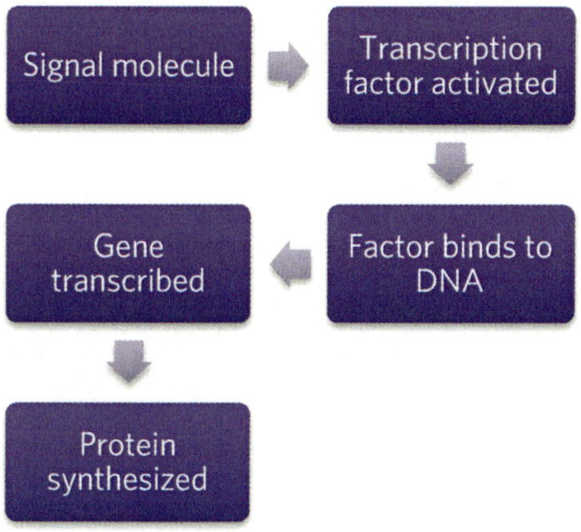

PLACING THE CALL

LOCAL OR LONG DISTANCE?

Chemicals deliver messages to nearby or distant cells. Cells secrete local regulators such as growth factors into the extracellular fluid. Long-range signals such as hormones travel through the bloodstream to distant tissues.

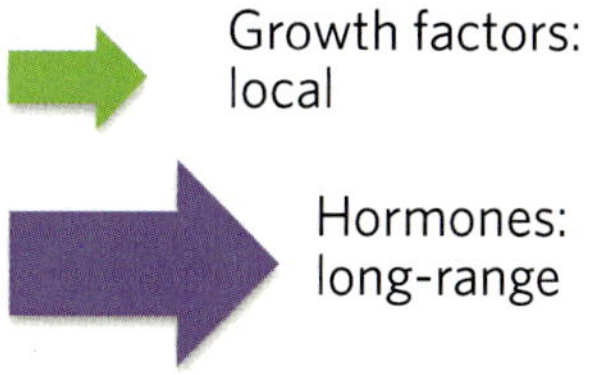

TURNING UP THE VOLUME

Cells respond to chemical signals with lightning speed thanks to small, water-soluble molecules called second messengers. These proteins or ions spread quickly through the cytoplasm, activating kinases and transmitting the signal. Cyclic AMP (cAMP) and inositol triphosphate (IP3) are two second messengers.

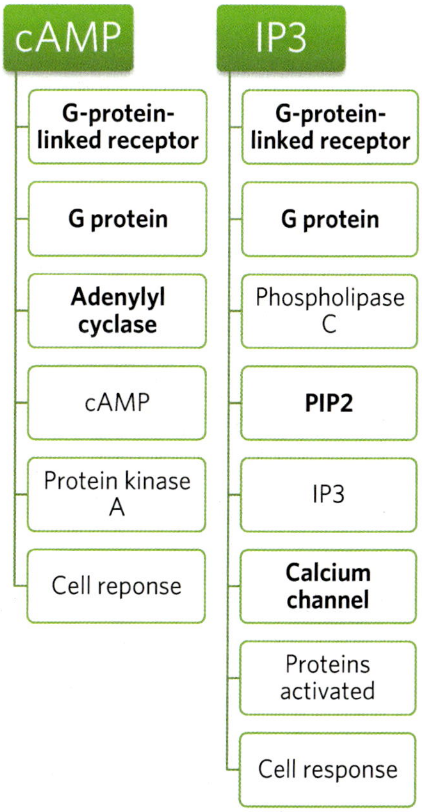

Note: Bolding indicates plasma membrane bound proteins or other compounds. Calcium channel is found in the ER's membrane.

CRAM QUIZ
Cell Signaling

QUESTION 1

Paracrine signaling in animals
(A) is a type of local signaling
(B) is a type of long-distance signaling
(C) targets single cells
(D) produces a weak response
(E) is less understood than paracrine signaling in plants

QUESTION 2

G-protein-linked receptors are not
(A) embedded in the plasma membrane
(B) important to the activation of certain enzymes
(C) activated by other molecules
(D) similar in structure
(E) composed of six alpha helices

QUESTION 3

Which step in the activation of tyrosine kinase is incorrect?
(A) A signal molecule attaches to the receptor.
(B) The receptor joins with its monomer to form a dimmer.
(C) The dimer releases its signal molecules.
(D) The dimer is phosphorylated.
(E) The active receptor binds to specific proteins.

QUESTION 4

A neurotransmitter is also called a ligand because it
(A) causes synaptic transmission
(B) binds to an ion channel
(C) binds to a G-protein receptor
(D) closes a membrane channel
(E) transmits signals quickly

QUESTION 5

Protein phosphatases are important to signal transduction because they
(A) phosphorylate other molecules
(B) transmit a signal
(C) activate protein kinases
(D) dephosphorylate proteins
(E) are always in greater concentrations than kinases

QUESTION 6

Epinephrine stimulates the breakdown of glycogen by
(A) inactivating protein kinases
(B) binding to a tyrosine kinase receptor
(C) acting as a second messenger
(D) binding to a G-protein-linked receptor
(E) opening an ion channel

QUESTION 7

A possible function of scaffolding proteins is to
(A) maintain cell shape
(B) interact with the nuclear envelope
(C) activate enzymes
(D) remain attached to plasma membrane proteins
(E) carry many protein kinases

QUESTION 8

Calcium concentration is higher in the
(A) nucleus
(B) nucleus and the extracellular fluid
(C) cytoplasm and the extracellular fluid
(D) cytoplasm and the ER
(E) ER and mitochondria

ANSWERS

1. A
2. E
3. C
4. B
5. D
6. D
7. E
8. E

CELLS
Cellular Respiration

FOOD FOR THOUGHT

BREAKING DOWN RESPIRATION

Three pathways break down glucose (when oxygen is present) to produce ATP, the molecule that powers cell functions: glycolysis, citric acid cycle, and oxidative phosphorylation.

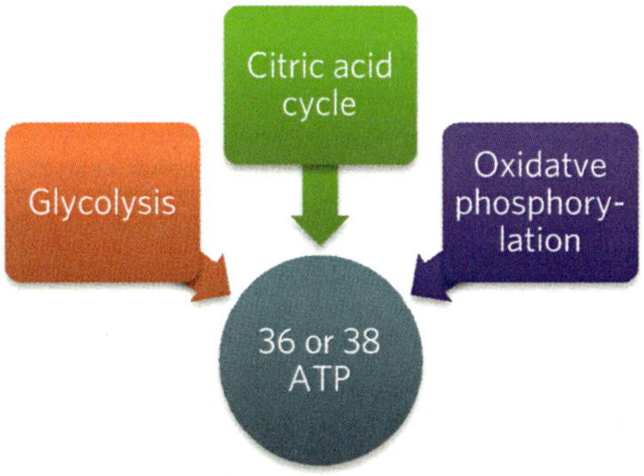

ENERGY FLOW

Cells release the energy stored in glucose and other molecules in several steps instead of all at once to control the reaction and maximize ATP production.

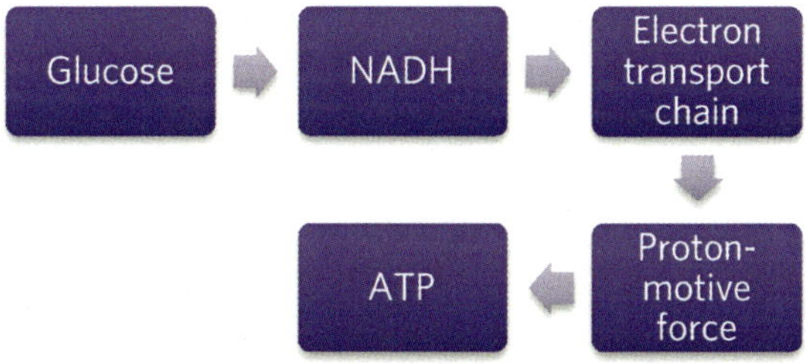

OXIDATIVE OR SUBSTRATE LEVEL?

Phosphorylation donates a phosphate group to a molecule such as ADP, which then becomes ATP. Breaking one of ATP's phosphate bonds releases energy, which fuels cellular processes. Donor phosphates come from two sources: a substrate molecule and inorganic phosphate.

Oxidative	Substrate level
Energy released during electron transport phosphorylates ADP with an inorganic phosphate.	Enzyme transfers phosphate from substrate to ADP
Produces 90% ATP	Produces a few ATP molecules
Chemiosmosis	Glycolysis, citric acid cycle, fermentation

SCORECARD: PROCESSES

GLYCOLYSIS

Main objective: Produce pyruvate to drive citric acid cycle and NADH for oxidative phosphorylation.

What Goes In	What Comes Out
1 glucose	2 pyruvate 2 water
2 ATP	4 ATP
2 NAD^+ 4 e^- 4 H^+	2 NADH 2 H^+

CITRIC ACID CYCLE

Main objective: Produce NADH and $FADH_2$ to drive oxidative phosphorylation.

What Goes In	What Comes Out
2 pyruvate 2 CoA	5 CO_2 2 CoA
	2 ATP
8 NAD^+ 2 FAD	8 NADH 8 H^+ 2 $FADH_2$

OXIDATIVE PHOSPHORYLATION

Main objective: MAKE ATP!

What Goes In	What Comes Out
10 NADH 10 H^+ 2 FADH	8 NADH 8 H^+ 2 $FADH_2$

CRAM QUIZ
Cellular Respiration

QUESTION 1

Redox reactions involve the transfer of electrons among reactants. A compound is reduced when it

(A) receives electrons
(B) donates electrons
(C) transfers electrons
(D) becomes the reducing agent
(E) no longer shares electrons in a covalent bond

QUESTION 2

The citric acid cycle gets its name from its first product, citrate. The molecule formed after the removal of CoA is

(A) fumarate
(B) oxaloacetate
(C) malate
(D) succinate
(E) isocitrate

QUESTION 3

Which functional group becomes unstable with the attachment of coenzyme A to an organic compound?

(A) carboxyl
(B) ketone
(C) phosphate
(D) acetone
(E) acetyl

QUESTION 4

Cytochromes function in electron transport by accepting and donating electrons. What atom performs this function?

(A) hydrogen
(B) iron
(C) nitrogen
(D) oxygen
(E) zinc

QUESTION 5

In cellular respiration, NAD+ functions as a(n)

(A) reducing agent
(B) proton donor
(C) oxidizing agent
(D) electron donor
(E) hydroxide ion acceptor

QUESTION 6

Fermentation is the anaerobic (without oxygen) catabolism of organic compounds that generates ATP as long as what molecule is recycled?

(A) NADH
(B) FADH
(C) FAD+
(D) NAD+
(E) pyruvate

QUESTION 7

The two types of fermentation differ based on

(A) their initial reactants
(B) their initial products
(C) their electron acceptors
(D) the number of organic molecules produced
(E) the number of ATP molecules expended

QUESTION 8

Fat molecules store energy in their fatty acid tails. After beta oxidation, these molecules enter

(A) glycolysis
(B) citric acid cycle
(C) chemiosmosis
(D) the Calvin cycle
(E) oxidative phosphorylation

ANSWERS

1. A
2. D
3. E
4. B
5. C
6. D
7. B
8. B

CELLS
Cell Cycle

IT'S JUST A PHASE

GROWING UP

A cell passes through two stages in its lifetime: interphase (growth period) and mitosis (cell division), each of which has its own phases.

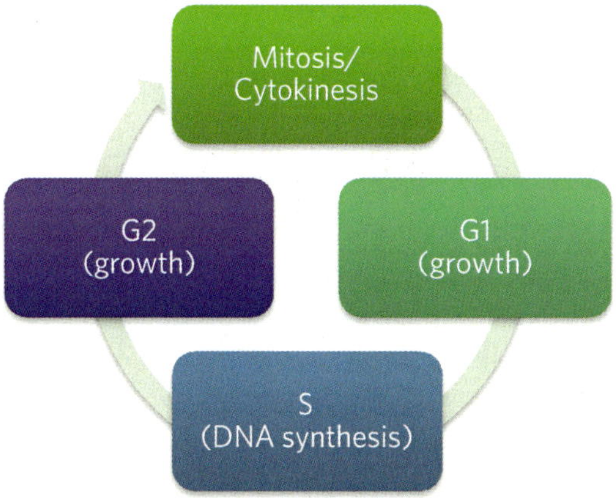

CLOSER LOOK: MITOSIS

Phase	*What's Happening*
G2	Chromatin and centrosomes visible
Prophase	Chromosomes condense, and the mitotic spindle forms. Nucleoli disappear. Centrosomes separate.
Prometaphase	Microtubules from centrosomes attach to chromatids' kinetochores. Nuclear envelope fragments.
Metaphase	Chromosomes line up along the metaphase plate.
Anaphase	Sister chromatids migrate to opposite poles.
Telophase	Nuclear envelopes reform around daughter nuclei and less condensed chromosomes.
Cytokinesis	Cytoplasm divides. Two daughter cells form.

THE RIGHT TIMING

CHECK IT OUT

A group of cyclical molecules called kinases controls the cell cycle with stop-and-go signals at three major checkpoints.

- The G1 checkpoint is the most important regulatory step in the cell cycle; kinases signal the cell to continue through interphase and into mitosis or send the cell into a non-dividing state (G0 phase)
- Cyclin activates kinases by forming cyclin-dependent kinases (Cdk); M-phase-promoting factor (MPF) is one type of Cdk
- Generally, cells have built-in stop signals; Cdks override stop signals and push the cell into the next phase

MPF AND G2 CHECKPOINT

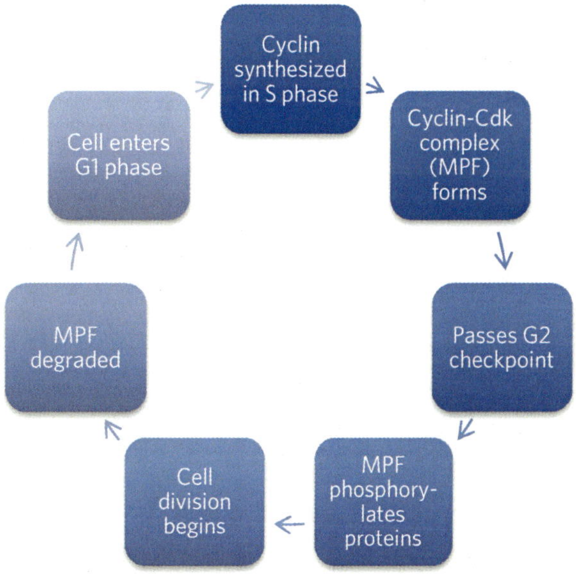

DID YOU KNOW?

Internal and external signals trigger cell growth and division.

- **Internal:** Cells pass the M checkpoint after a protein secreted by the kinetochores is inactivated once the cells attach sister chromatids to the mitotic spindle
- **External:** Chemicals secreted by cells called growth factors stimulate division

CRAM QUIZ
Cell Cycle

QUESTION 1

Which phase of the cell cycle varies the most in different cell types?

(A) M phase
(B) S phase
(C) G2 phase
(D) G1 phase
(E) interphase

QUESTION 2

Contractile ring formation in animals differs from cell plate formation in plants in that

(A) the animal cell expands
(B) a cleavage furrow forms in animal cells
(C) the cell plate builds from the cell wall inward and vice versa in animals
(D) microfilaments form cell plates and vesicles form contractile rings
(E) the cell plate contracts the plant cell

QUESTION 3

The asters appear at the end of which phase of mitosis?

(A) prophase
(B) metaphase
(C) G2 of interphase
(D) prometaphase
(E) telophase

QUESTION 4

Sister chromatids are connected at what structure?

(A) centrosome
(B) centromere
(C) kinetochore
(D) centrioles
(E) mitotic spindle

QUESTION 5

Microtubules that do not attach to the kinetochores of sister chromatids

(A) walk chromosomes to opposite poles
(B) shorten and pull chromosomes to opposite poles
(C) stretch the cell lengthwise
(D) move closer to the microtubules attached to the kinetochores
(E) disappear after metaphase

QUESTION 6

Which experiment would determine whether a kinetochore shortens at the spindle or kinetochore ends?

(A) observing microtubule movement during metaphase
(B) dyeing sister chromatids during early anaphase
(C) dyeing microtubules during early anaphase and measuring their lengths at telophase
(D) removing dye from a segment of microtubules and measuring the lengths on either side of the mark during anaphase
(E) cutting microtubules near the spindle and measuring the lengths on each side of the cut

QUESTION 7

What structure does not replicate during mitosis?

(A) centrosomes
(B) chromosomes
(C) centrioles
(D) chromatin
(E) nuclear envelope

QUESTION 8

What happens to the cell if it doesn't receive the go-ahead signal at the "restriction point"?

(A) It passes into a nondividing state.
(B) It moves into G1 phase.
(C) It undergoes apoptosis.
(D) It continues into the S phase.
(E) Cyclin accumulates in the cell.

ANSWERS

1. D
2. D
3. A
4. B
5. C
6. E
7. B
8. A

CELLS
Viruses

PERFECT PARASITES

SIMPLE STRUCTURE, BIG EFFECT

Viruses are simple infectious agents that force a host cell to churn out more viruses. They are similar to cells in that they contain genetic material, are surrounded by a membrane or protein coat (or both), and replicate.

PHAGE STRUCTURE

Viruses come in many forms. A well-known example of a virus is the one that infects *E. coli* bacteria.

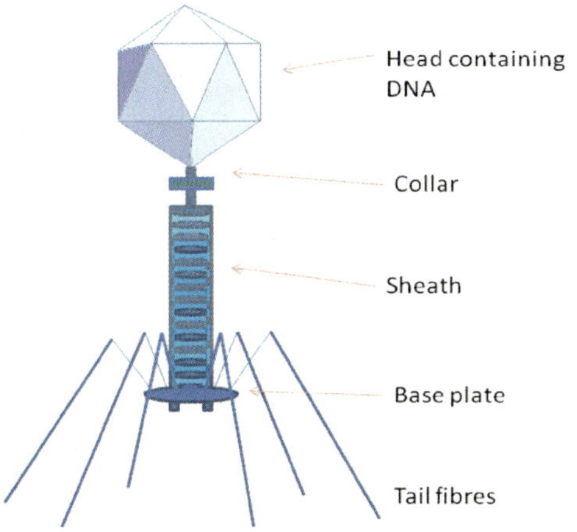

HARD TO CRACK

An outer casing (capsid) protects a virus and may be surrounded by a membrane (envelope).

	Capsid	*Envelope*
Derived from	Virus	Host
Composition	Protein subunits (capsomeres)	Host membranes Viral proteins
Function	Protect nucleic acids	Help with infection
Shapes	Rod, polyhedral, icosahedron	Spherical
Examples	T-phages, adenovirus	Influenza, retrovirus

QUICK REFERENCE: VIRAL REPRODUCTION

1. Virus enters host cell.
2. Host replicates, transcribes, and translates viral genome.
3. Virus assembles and exits cell.

HIJACKING A HOST

TO KILL OR NOT TO KILL?

Bacteriophages (viruses that infect bacteria) can kill their hosts by quickly lysing the cell, or they can harmlessly integrate into the host's genome (and kill the host later).

VIRAL REPRODUCTION: DEADLY OR NOT

At any time, a phage can exit the lysogenic cycle, enter the lytic cycle, and kill the host.

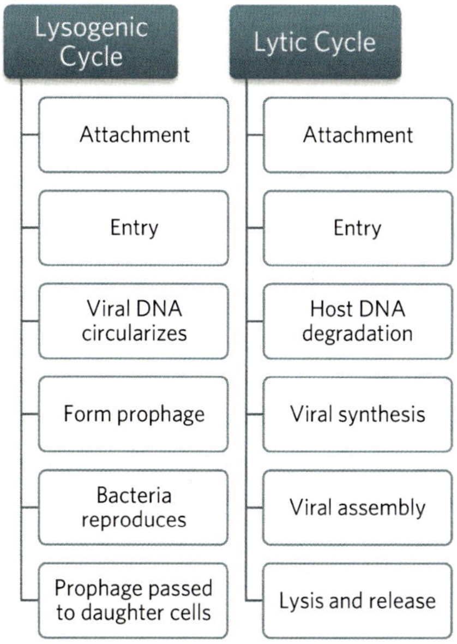

HARD TO KILL: A CLOSER LOOK AT HIV

HIV is a retrovirus with a complex reproductive cycle that damages the immune system by binding to white blood cells.

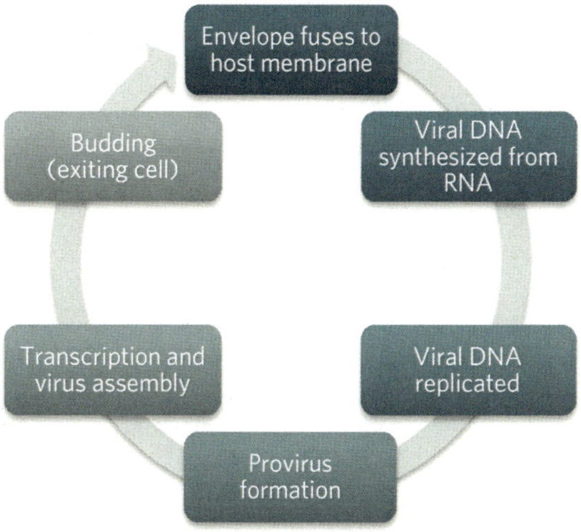

CRAM QUIZ
Viruses

QUESTION 1

Viroids are

(A) harmless derivatives of viruses
(B) infectious proteins
(C) RNA molecules that infect plants
(D) DNA molecules that infect bacteria
(E) viruses without an envelope

QUESTION 2

The difference between a prophage and a provirus is that a

(A) provirus never leaves the host's genome.
(B) prophage never leaves the host's genome.
(C) prophage only infects bacteria.
(D) provirus is synthesized by reverse transcriptase.
(E) provirus uses restriction enzymes to enter a host's genome.

QUESTION 3

What protective measures does a bacterium have against viral infection in addition to restriction enzymes that cut up foreign DNA?

(A) hard protein coats that viruses can't enter
(B) enzymes that prevent viral DNA replication
(C) a jelly-like substance on the membrane that prevents attachment
(D) unfamiliar receptor sites on membranes
(E) the breakdown of foreign proteins by enzymes

QUESTION 4

What causes many of the symptoms associated with viral infections?

(A) lysis
(B) the immune system
(C) viral reproduction
(D) vaccines
(E) phages

QUESTION 5

Viruses possibly originated from

(A) protobionts
(B) early forms of cells
(C) plasmids
(D) bacteria
(E) transposable nucleic acids

QUESTION 6

What usually triggers a virus to exit the lysogenic cycle and enter the lytic cycle?

(A) environmental factors
(B) production of certain viral proteins
(C) bacterial division
(D) excision of a prophage
(E) production of toxins

QUESTION 7

Assembly of new viruses is carried out

(A) by host structures
(B) within a vesicle
(C) with the assistance of viral enzymes
(D) spontaneously
(E) after the production of a particular enzyme

QUESTION 8

How many different classes of animal viruses exist?

(A) two
(B) three
(C) four
(D) five
(E) six

ANSWERS

1. C
2. A
3. D
4. B
5. E
6. A
7. D
8. E

HEREDITY AND GENETICS
Sexual Life Cycles and Meiosis

THE CYCLE OF LIFE

THE NEXT GENERATION

Organisms pass traits on to the next generation with gametes—sex cells that contain genetic material. During fertilization, gametes of mating individuals combine to form a new organism.

CHROMOSOMAL CONNECTION

- **Chromosomes** are the source of heredity
- All cells with the exception of gametes contain two sets of chromosomes (**diploid**, *2n*)
- **Homologues** are pairs of chromosomes with genes that control the same traits (for example, eye color)
- **Mitosis** produces cells with the same number of chromosomes as the parent
- **Meiosis** produces gametes with half the number of chromosomes as the parent (**haploid**, *n*)
- During **fertilization**, haploid gametes fuse to form a diploid nucleus

SEXUAL LIFE CYCLES

Sexual life cycles differ between organisms based on the timing of meiosis and fertilization. Other differences include:

Animals produce diploid multicellular organisms.

Plants/some algae produce sporophytes (diploid), spores (haploid), and gametophytes (haploid).

Most fungi/some protists produce diploid multicellular organisms.

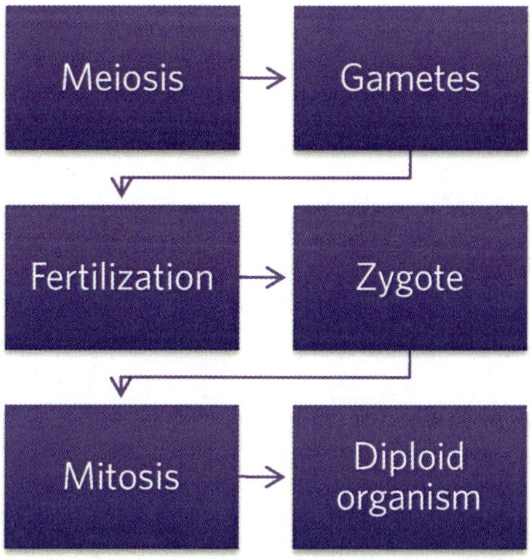

Note: Cells in the haploid stage of plants, algae, fungi, and protists also undergo mitosis.

NOT HALF BAD

MAKING GAMETES

Meiosis cuts the number of chromosomes in half so that when gametes fuse during fertilization, the offspring has the same genome as the parents.

OVERVIEW: MEIOSIS

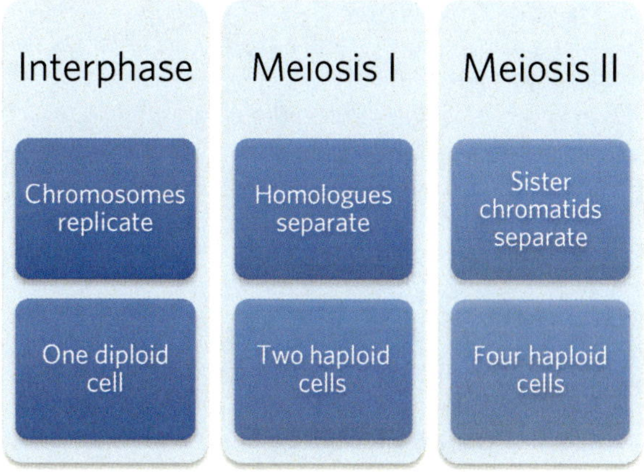

A CLOSER LOOK AT MEIOSIS I/MEIOSIS II

Stage	*What Happens*
Prophase I Prophase II	Nonsister chromatids undergo crossing over Synapsis occurs Nuclear envelope disintegrates Spindle forms Spindle forms Homologues/sister chromatids move to metaphase plate
Metaphase I Metaphase II	Kinetochores of homologues/sister chromatids attach to microtubules
Anaphase I Anaphase II	Homologues/sister chromatids separate and move to opposite poles
Telophase I and II and Cytokinesis	Cytokinesis Nuclei reform* Chromosomes decondense*

**This occurs in Telophase I in some species.*

CRAM QUIZ
Sexual Life Cycles and Meiosis

QUESTION 1

Plants and some algae undergo alternation of generations because

(A) their haploid state divides mitotically to form a multicellular organism
(B) some organisms reproduce asexually and others reproduce sexually
(C) they produce spores and a diploid organism called a sporophyte
(D) they produce haploid and diploid multicellular organisms
(E) the diploid zygote undergoes mitosis

QUESTION 2

What are the steps of tetrad formation in proper order? A) Crossing over occurs. B) Synaptonemal complex disassembles. C) Homologues pair up gene by gene. D) Synaptonemal complex forms.

(A) C, D, B, A
(B) D, B, A, C
(C) C, D, A, B
(D) A, D, C, B
(E) B, C, A, D

QUESTION 3

The differences between meiosis and mitosis include

(A) the genetic composition of daughter cells
(B) DNA is not replicated in meiosis
(C) synapsis occurs at different stages
(D) meiosis involves four times as many divisions as mitosis
(E) mitosis produces genetic variability

QUESTION 4

Meiosis I is referred to as reductional division because it

(A) reduces the number of chromosome sets by half
(B) separates sister chromatids into daughter cells
(C) is when crossing over occurs
(D) reduces the number of chromosomes by half
(E) is when independent assortment of homologues occurs

QUESTION 5

Which of the following processes encourages genetic diversity?

(A) synapsis, separation of homologues, and crossing over
(B) crossing over, independent assortment, and tetrad formation
(C) sexual reproduction, random fertilization, and synapsis
(D) random fertilization, crossing over, and independent assortment
(E) gamete formation, sexual reproduction, and crossing over

QUESTION 6

The independent assortment of homologues in meiosis produces daughter cells with what probability of containing all maternal genes (without accounting for crossing over)?

(A) 0%
(B) 25%
(C) 50%
(D) 75%
(E) 100%

QUESTION 7

Asexual reproduction is important for

(A) genetic diversity
(B) gamete formation
(C) clonal production
(D) animal reproduction
(E) exact copying of genetic material of parent

QUESTION 8

The chiasma

(A) holds chromosomes together for crossing over
(B) is the same as a tetrad
(C) is formed in prophase II
(D) occurs between sister chromatids
(E) is an X-shaped region of homologues

ANSWERS

1. D
2. C
3. A
4. A
5. D
6. B
7. C
8. E

HEREDITY AND GENETICS
Laws of Inheritance

MAGNIFICENT MONK

THE GARDEN OF GENES

An Austrian Monk named Mendel fathered modern genetics by crossing plants in an abbey garden. He chose common pea plants for his experiments because

1. They come in several varieties.
2. Their breeding is easy to control.
3. Many plant traits are "either-or." (For example, a plant has either purple or white flowers.)

CREATE YOUR OWN MENDELIAN EXPERIMENT!

WHAT YOU NEED

Scissor, paint brush, and two true-breeding plants with different traits*

***Hey, what's this?**

When true-breeding plants self-pollinate, their offspring have the same traits as the parents. Traits are the variations of certain characteristics like flower color or pod type. Flowers come in two varieties: white and purple.

WHAT YOU DO

1. Remove stamens from one plant with scissors, exposing the carpel.
2. Use the paint brush to transfer pollen from the stamens of the other plant to the exposed carpel.
3. When the pods mature, plant seeds from the pod.
4. Self-pollinate the offspring.
5. Observe the traits of the second generation.

WHAT HAPPENS

Parental generation (P1)

Purple flower (PP)* — White flower (pp)

↓

First generation offspring (F1)

All purple flowers (Pp)

↓

Second generation offspring (F2)

3 purple flowers (1 PP, 2Pp) — 1 white flower (pp)

**P indicates the dominant gene and p the recessive gene. Hybrids (Pp) display the dominant trait.*

GENETIC LAWS

WHAT DOES IT MEAN?

After repeating his experiments many times, Mendel developed the laws of heredity from his observations.

MENDEL'S CONCLUSIONS WITH A MODERN TWIST

- Alternate gene forms called alleles explain variations in hereditable characteristics.
- Each parent passes one allele for a gene to its offspring.
- An allele may be dominant (physically visible) or recessive (masked by a dominant allele).
- Alleles separate into different gametes independently of other allele pairs.

EXCEPTIONS TO MENDEL'S RULES

Incomplete Dominance

Some alleles are not completely dominant over others and may fall more in the middle. (Hybrid snapdragons, for example, aren't red or white, but pink.)

Pleiotropy

One gene affects phenotype in many ways.

Epistasis

A gene at one locus affects the expression of a gene at another locus.

Polygenic Inheritance

Several genes have an additive effect on one characteristic.

QUICK REFERENCE

Dominant	Visible trait
Recessive	Unnoticeable trait
Genotype	Genetic makeup
Phenotype	Distinguishable trait
Heterozygous	Having different alleles for a gene
Homozygous	Having same alleles for a gene
Locus	The location of a gene on the chromosome
Epistasis	When crossing two heterozygous parents yields three different phenotypes

CRAM QUIZ
Laws of Inheritance

QUESTION 1

The F2 generation offspring of organisms affected by incomplete dominance occurs in what ratio?

(A) 1:4:1
(B) 2:3:2
(C) 3:1
(D) 1:2:1
(E) 1:3:1

QUESTION 2

Segregation of gametes and random fertilization affect offspring genotypes. What is the probability that an offspring of two heterozygotes will be homozygous recessive?

(A) 1/8
(B) 1/4
(C) 1/2
(D) 1
(E) 0

QUESTION 3

Mendel developed the law of independent assortment by

(A) self-pollinating dihybrids
(B) crossing P and F1 generations
(C) crossing two monohybrids
(D) predicting that the phenotypic ration for the F2 generation is 3 to 1
(E) crossing plants with many traits

QUESTION 4

Crossing two heterozygous black mice produces black, brown, and albino progeny in what ratio, respectively?

(A) 10/16, 2/16, 4/16
(B) 11/16, 3/16, 3/16
(C) 9/16, 3/16, 4/16
(D) 12/16, 3/16, 2/16
(E) 9/16, 4/16, 3/16

QUESTION 5

Determination of blood type is an example of

(A) epistasis
(B) incomplete dominance
(C) polygenic inheritance
(D) pleiotropy
(E) multiple alleles

QUESTION 6

The norm of reaction is

(A) the genotypic diversity of living organisms
(B) the range of phenotypes for every genotype
(C) the effect of the environment on blood type
(D) broadest for pleiotropic alleles
(E) uncommon in animals

QUESTION 7

An example of a multifactorial disorder is

(A) Huntington's disease
(B) achondroplasia
(C) schizophrenia
(D) cystic fibrosis
(E) Tay Sachs disease

QUESTION 8

In a testcross, an organism with a recessive phenotype is crossed with

(A) another recessive organism
(B) an organism with a known genotype
(C) an heterozygous organism
(D) an organism of unknown genotype
(E) an organism of known phenotype

ANSWERS

1. D
2. B
3. A
4. C
5. E
6. B
7. C
8. D

HEREDITY AND GENETICS
Chromosomes and Genes

GARDEN PEAS AND FRUIT FLIES

FROM MENDEL TO MORGAN

Morgan proved Mendel's "heritable factors" were actually genes located on chromosomes by experimenting with the fruit fly *Drosophila*.

CHROMOSOMES ARE THE KEY

Taking a closer look at Mendel's experiments with true-breeding yellow-round peas (YYRR) and green-wrinkled peas (yyrr) shows that during gamete formation...

1. The two alleles for pea color and type separate. **(LAW OF SEGREGATION)**
2. Alleles for pea color and type (which are genes on nonhomologous chromosomes) separate independently of each other. **(LAW OF INDEPENDENT ASSORTMENT)**

Genotypes of F1 Generation Gametes

1/4 YR	1/4 yr
1/4 Yr	1/4 yR

9:3:3:1*

**The phenotypic ratio of F2 generation offspring: 9 yellow round, 3 yellow green, 3 yellow wrinkled, 1 green wrinkled.*

WHEN THINGS BREAK DOWN

Occasionally, a chromosome breaks during crossing over or doesn't separate or move properly during meiosis (nondisjunction). This error can lead to changes in chromosome number or structure of the gametes. In most cases, these alternations are harmful to the organism.

Aneuploidy	Abnormal number of chromosomes
Deletion	Segment of DNA removed
Duplication	Segment of DNA removed
Inversion	Segment of DNA inverted
Translocation	Segment of DNA moved to nonhomologous chromosome

THE BATTLE OF THE SEXES

LINKING EYE COLOR TO SEX

Morgan bred thousands of red-eyed fruit flies in search of a rare white-eyed fly. He observed Mendel's 3 to 1 ratio in the offspring with a surprising twist: all white-eyed flies were male.

MORGAN'S RESULTS

P1 generation
Red-eyed females X White-eyed males

↓

F1 generation
Red-eyed females X Red-eyed males

↓

F2 generation
Red-eyed females, Red-eyed males, White-eyed males
3 red eyes to 1 white eye

SURPRISING CONCLUSIONS

In general, females have a homologous pair of X chromosomes (XX). Males have one X and one Y chromosome (XY). The gene for eye color in flies is carried on the X chromosome (it's a **sex-linked gene**). Mothers pass sex-linked genes on to sons and daughters; fathers pass them on to daughters.

NOT TO BE CONFUSED WITH . . .

Morgan discovered an exception to Mendel's law of independent assortment: when genes are located on the same chromosomes, they may be inherited together (they are **linked**).

FINDING THE FREQUENCY

Recombination frequencies show whether genes are inherited together (see below) or located on different chromosomes (50% frequency).

$$\frac{\text{Recombinants}}{\text{Offspring}} = \text{Recombination frequency}$$

Recombinants are offspring that are different from their parents.

DRAW A MAP

A linkage map assumes that the probability of crossing over is proportional to the distance between two linked genes. One unit on a linkage map equals 1% recombination frequency.

CRAM QUIZ
Chromosomes and Genes

QUESTION 1

Three genes (A, B, and C) are located on the same chromosomes. A and B have recombination frequencies of 18%. A and C have frequencies of 10%. What is the order of the genes?

(A) A, B, C
(B) A, C, B
(C) B, A, C
(D) C, A, B
(E) B, C, A

QUESTION 2

How does the distance between two genes affect crossing over?

(A) The closer two genes are, the higher the probability that crossing over will occur.
(B) The closer two genes are, the more distant their loci.
(C) The farther apart two genes are, the higher the probability that crossing over will occur.
(D) Crossing over doesn't occur between genes on the same chromosomes.
(E) Crossing over doesn't relate to the distance between genes.

QUESTION 3

The inactivation of the X chromosome (Barr body) in females is due to the

(A) condensing of the chromosomes
(B) removal of chromosomes from nucleus
(C) inability to express all genes on the chromosome
(D) a non-random process
(E) double dosing of genes in cells of females

QUESTION 4

Genomic imprinting causes changes in phenotype based on whether an allele is inherited from the mother or father. This phenomenon is due to

(A) a process that occurs in most genes
(B) activation of both copies of a gene
(C) a process that occurs after embryonic development
(D) inactivation of genes
(E) methylation of certain nucleotides

QUESTION 5

Organelle genes are passed on to offspring

(A) in the same way as nuclear genes
(B) by being transferred to the nucleus during replication
(C) by way of the male gamete
(D) by replication of organelles in egg cytoplasm
(E) by being incorporated into the genome

QUESTION 6

Crossing over occurs between

(A) sister chromatids in prophase II
(B) homologues in prophase I
(C) homologues in metaphase II
(D) sister chromatids in metaphase I
(E) homologues in prophase II

QUESTION 7

Nondisjunction of chromosomes in meiosis I causes what alternations to chromosomal numbers (identified as n)?

(A) 2 gametes with n; 1 gamete with $n-1$; 1 gamete with $n+1$
(B) 2 gametes with n; 2 gametes with $n+1$
(C) 2 gametes with $n-1$; 2 gametes with $n+1$
(D) 2 gametes with n; 2 gametes with $n-1$
(E) does not affect chromosome number

QUESTION 8

Males express recessive sex-linked alleles if they

(A) are homozygous
(B) are heterozygous
(C) receive the allele from their father
(D) are hemizygous
(E) are not affected by sex-linked alleles

ANSWERS

1. B
2. C
3. A
4. E
5. D
6. B
7. C
8. D

HEREDITY AND GENETICS
DNA

TWISTED PAIR

THE DOUBLE HELIX

Chromosomes are made up of DNA, a helical molecule containing the genetic material that is passed on from one generation to the next.

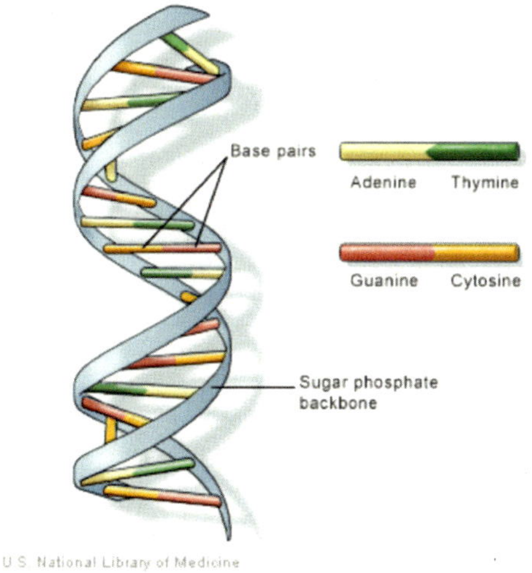

Source: Genetics Home Reference (published August 14, 2009)

SPECIAL STRUCTURE

- Hydrogen bonding holds nitrogenous bases (adenine, thymine, guanine, and cytosine) together in the center of the helix
- A sugar phosphate backbone runs along the outside
- Purines (A and G) only bind to pyrimidines (T and C) and vice versa, making DNA strands complementary: A always binds to T, and G always binds to C

DIFFERENT ENDS

A closer look at DNA's sugar-phosphate backbone reveals two different ends, the 5′ and 3′ ends. The molecue is oriented in opposite directions (antiparallel).

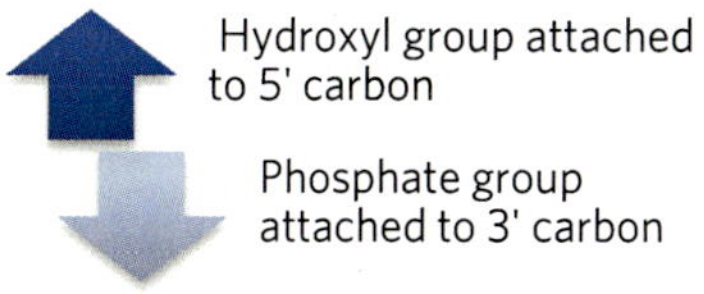

MAKIN' COPIES

DNA replication takes place at sites along the molecule called origins of replication. Proteins "unzip" DNA's double helix, creating replication bubbles.

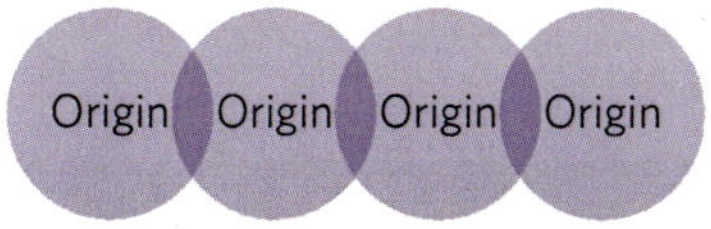

TWO IS BETTER THAN ONE

REPLICATION IN A NUTSHELL

DNA replication is semi-conservative—each parent DNA strand serves as a template for a new strand.

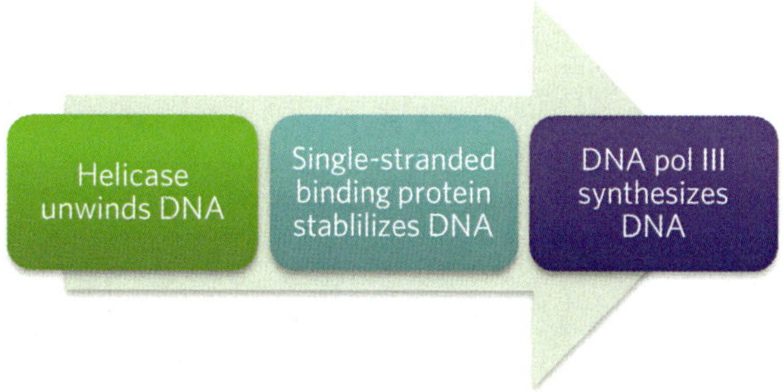

WARNING!
Only replicate DNA in the 5′ to 3′ direction! (If you're DNA polymerase III, that is.)

DON'T LAG BEHIND

DNA replication of the 3′ leading strand is continuous. The lagging strand replicates in fragments, but the overall direction of replication of the new strand is 5′ to 3′.

1. Primase synthesizes primer for fragment that will be replicated (Okazaki fragment).
2. DNA pol III binds to primer and adds nucleotides in the 5′-to-3′ direction.
3. DNA pol I removes primer.
4. DNA ligase binds the fragments together.

FIXING MISTAKES

DNA polymerases proofread their work as they replicate the DNA molecules. If they miss a mistake, nuclease cuts out mismatched nucleotides.

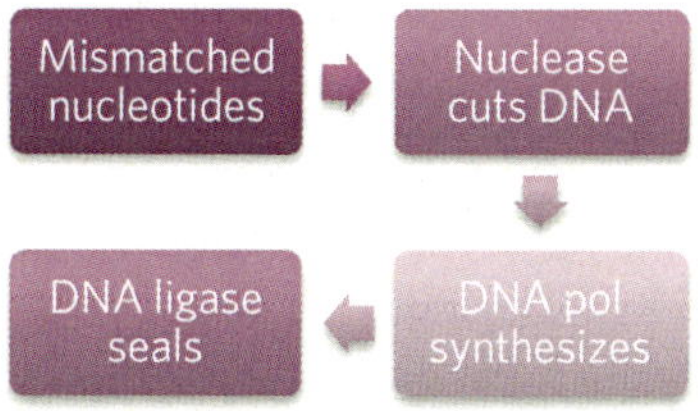

TELOMERES AND DNA PROTECTION

The last RNA primer attached to the lagging strand of a DNA molecule is never replaced with DNA. To prevent DNA from shortening every time it's replicated, repetitive, noncoding sequences called telomeres are found at the end of DNA molecules.

CRAM QUIZ
DNA

QUESTION 1

The lagging strand gets its name from the

(A) enzyme that synthesizes it
(B) added step of primase synthesizing a primer before DNA polymerase starts replication
(C) 3′ end
(D) time it takes the replication fork to expose enough DNA replication to begin
(E) extra time it takes to synthesize the strand in fragments

QUESTION 2

X-ray crystallograph demonstrates the helical structure of DNA and shows

(A) the helix's width and spacing of nitrogenous bases
(B) how long the DNA strands are
(C) how nitrogenous bases are paired
(D) the percentages of nitrogenous bases in a DNA molecule
(E) covalent bonding in the sugar-phosphate backbone

QUESTION 3

Which of the following enzymes does not function in DNA replication?

(A) helicase
(B) nuclease
(C) topoisomerase
(D) single-stranded binding protein
(E) ligase

QUESTION 4

Environment phenomena such as radiation and cellular conditions can damage DNA. The team of enzymes that function in DNA repair are

(A) helicase, topoisomerase, ligase
(B) primase, nuclease, ligase
(C) DNA polymerase, nuclease, helicase
(D) DNA polymerase, nuclease, ligase
(E) ligase, DNA polymerase, helicase

QUESTION 5

Telomerase lengthens telomeres in what type of cells?

(A) eukaryotic cells
(B) somatic cells
(C) germ cells
(D) prokaryotic cells
(E) stem cells

QUESTION 6

According to the dispersive model of DNA replication,

(A) some daughter strands are completely new and some contain one parent and one new strand
(B) some daughter strands are completely new and some are from parent strands
(C) some daughter strands are from parent strands and some contain one parent and one new strand
(D) some daughter strands are completely new and some contain a mixture of new and parent DNA
(E) all daughter strands are a mixture of new and parent DNA

QUESTION 7

DNA polymerase III adds which of the following to the growing strand of a DNA molecule?

(A) nucleoside triphosphate
(B) nucleotide triphosphate
(C) nucleoside
(D) nucleotide
(E) sugar and a nitrogenous base

QUESTION 8

DNA replication starts with the attachment of a primer that is made of

(A) RNA
(B) DNA
(C) RNA or DNA
(D) protein
(E) RNA or protein

ANSWERS

1. D
2. A
3. B
4. D
5. C
6. E
7. A
8. C

HEREDITY AND GENETICS
RNA

THE PERFECT GO-BETWEEN

BRIDGING THE GAP

RNA passes along the genetic information found in DNA by directing protein synthesis (translation). DNA serves as the template for RNA synthesis (transcription).

SIMILAR STRUCTURES

DNA and RNA have similar structures that facilitate transcription, with a few notable differences.

DNA	RNA
4 types of nucleotides	4 types of nucleotides
Nitrogenous bases (**thymine**)	Nitrogenous bases (**uracil**)
Deoxyribose	**Ribose**
Double stranded	**Single** stranded (usually)

TRANSCRIPTION: STEP BY STEP

Transcription is the second step in the transfer of heritable information from DNA to cellular functions.

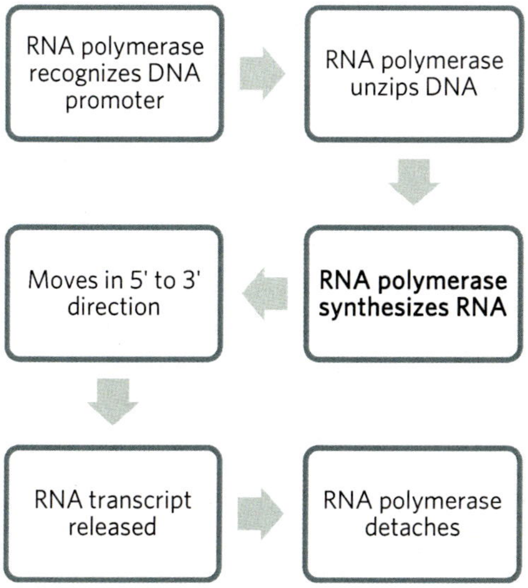

WORKING OVERTIME

NOT DONE YET

Eukaryotes add a few more steps to transcription: RNA polymerase II can't bind to DNA without the help of proteins called **transcription factors**. In addition to the three stages of RNA synthesis (initiation, elongation, and termination), eukaryotes don't use their RNA until it has been processed.

CUT AND PASTE

RNA processing called **splicing** removes large portions of a newly formed RNA molecule (exons), leaving regions that are used in protein synthesis (introns).

SPLICEOSOMES AND SNRPS

Processing of messenger RNA is carried out in a complex of proteins called a spliceosome, which includes RNA/protein particles (snRNPs) that recognize the ends of introns.

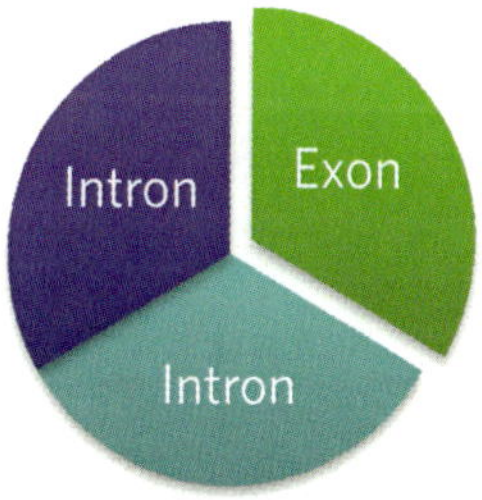

VERY VERSATILE

RNA serves many functions in eukaryotic cells.

mRNA	Directs protein synthesis
tRNA	Assists mRNA in translation
rRNA	Enzyme; important to ribosome structure
Transcript	RNA before splicing
snRNA	Structural and enzymatic role in spliceosomes
SRP RNA	Recognizes signal molecules
snoRNA	Assists pre-rRNA processing
siRNA, miRNA	Involved in gene expression

CRAM QUIZ

RNA

QUESTION 1

How does transcription differ in prokaryotes and eukaryotes?

(A) Eukaryotes have one type of polymerase that performs many functions.
(B) A terminator signals the end of transcription in prokaryotes.
(C) Prokaryotes can't initiate transcription without transcription factors.
(D) Elongation in eukaryotes occurs in the 3′ to 5′ direction.
(E) RNA is immediately usable in eukaryotes.

QUESTION 2

A molecule that can catalyze the removal of introns from pre-mRNA is a

(A) ribozyme
(B) spliceosome
(C) snRNP
(D) polymerase
(E) nuclease

QUESTION 3

What is the functional significance of RNA splicing?

(A) Exons increase the frequency of crossing over.
(B) It facilitates intron shuffling.
(C) It keeps RNA from degrading with each replication.
(D) It keeps mRNA in the nucleus.
(E) It allows genes to encode more than one protein.

QUESTION 4

The initiation phase of transcription in eukaryotes proceeds in what order?

(A) RNA polymerase II recognizes the promoter and binds to transcription factors and then to DNA.
(B) Transcription factors unwind DNA; RNA polymerase II binds to DNA.
(C) RNA polymerase II binds to TATA box.
(D) Transcription factors bind to DNA; RNA polymerase II forms the transcription initiation complex.
(E) RNA polymerase II binds to a DNA primer.

QUESTION 5

During RNA processing, how are the ends of a pre-mRNA molecule modified?

(A) A poly-A tail is added to the 5′ end, and a 3′ cap is added to the 3′ end.
(B) A modified adenine base is added to the 5′ end, and guanine nucleotides are added to the 3′ end.
(C) A modified guanine base is added to the 5′ end, and adenine nucleotides are added to the 3′ end.
(D) A telomere is added to both 5′ and 3′ ends.
(E) A poly-A tail is added at the promoter, and a 5′ cap is added at the 5′ end.

QUESTION 6

Which RNA molecule is found in the nucleolus?

(A) SRP RNA
(B) siRNA
(C) snRNA
(D) snoRNA
(E) miRNA

QUESTION 7

The versatility of RNA can be attributed to its ability to bond with DNA and RNA, its ability to form three-dimensional structures, and its

(A) enzymatic functional group
(B) ability to form hydrogen bonds
(C) processing after transcription
(D) easy transfer from one part of the cell to another
(E) ability to form small, functional particles

QUESTION 8

Modification of the ends of RNA molecules during processing may

(A) keep RNA inside the nucleus
(B) prevent degradation by enzymes
(C) affect folding of the translated protein
(D) facilitate attachment of ribosomes to the 3′ end of RNA molecules
(E) affect codon and anti-codon pairing

ANSWERS

1. B
2. A
3. E
4. D
5. C
6. D
7. A
8. B

HEREDITY AND GENETICS
Protein Synthesis

LOST IN TRANSLATION

BUILD YOUR OWN PROTEIN

RNA directs protein synthesis by serving as a template during translation. But how can four nucleotides code for 20 different amino acids?

TRIPLE TROUBLE

The answer lies in a **triplet code**: three consecutive nucleotides that correspond to a specific amino acid. mRNA codons account for the versatility of genetic coding.

DNA	RNA	Amino acid
• ACC	• UGG	• Tryptophan

KEY PLAYERS

Several molecules and structures are unique to translation.

tRNA	•Transfer amino acids to ribosomes •Anticodons recognize mRNA codons
Ribosome	•Site of translation •Two subunits
Aminoacyl-tRNA synthetase	•Join amino acid to tRNA
rRNA	•Make up ribosomes

TRANSLATION: STEP BY STEP

1. Small ribosomal unit binds to mRNA.
2. tRNA initiator binds to mRNA start codon (AUG) at ribosome's P site.
3. Large ribosomal unit forms initiation complex with the help of initiation factors and GTP.
4. tRNA base pairs with mRNA in A site of ribosome.
5. rRNA catalyzes peptide bond formation between amino acids.
6. Ribosome moves tRNA (minus amino acid) from P to E site and releases it.
7. tRNA with growing polypeptide moves from A to P site.
8. Stop codon in A site causes release factor to attach.
9. Release factor releases polypeptide from tRNA.
10. Complex, including ribosomal units, disassembles.

CHANGING IT UP

MAGNIFICENT MODIFICAITON

Like RNA transcripts, newly made proteins undergo processing called post-translational modifications that alter function or target the protein for a specific location.

MODIFICATIONS

- Chaperon proteins assist in protein folding
- Attachment of lipids, sugars, or other molecules
- Cutting into two or more pieces
- Bringing two or more proteins together
- **Signal peptide** is added to secreted proteins or proteins sent to endomembrane system

FREE OR BOUND?

Ribosomes roaming freely in the cytosol make proteins that function in the cytosol. Ribosomes bound to the ER or nuclear envelope make proteins for the endomembrane system or secreted proteins.

ONE LAST POINT

DNA MUTATIONS

Mutagens are agents that damage DNA by causing mutations (changes in DNA structure). Even changing one base pair (point mutation) can have dramatic effects.

Base-pair substitutions	•Replace nucleotide and its partner •May alter proteins
Silent mutations	•Caused by base-pair substitutions •Same amino acids are translated •Have no effect on proteins
Missense mutations	•Caused by substitution mutations •May not code for right amino acid
Nonsense mutations	•Codon mutates into a stop codon •Creates shorter, nonfunctional protein
Insertions	•Addition of nucleotide pair •Alter reading frame (frameshift mutation) •Cause many problems (missense, nonsense, termination)
Deletions	•Loss of nucleotide pair •Alter reading frame (frameshift mutation) •Cause many problems (missense, nonsense, termination)

CRAM QUIZ
Protein Synthesis

QUESTION 1

Why does tRNA look like a cloverleaf in one dimension but L-shaped in three dimensions?

(A) The anticodon region and other nucleotides form hydrogen bonds.
(B) Stretches of complementary nucleotides form hydrogen bonds.
(C) A loop at the 3′ end folds back on itself due to hydrogen bonding.
(D) The anticodon folds back on itself due to hydrogen bonding.
(E) tRNA has neither of these structures.

QUESTION 2

There are how many different aminoacyl-tRNA synthetase molecules?

(A) 4
(B) 1
(C) 20
(D) 3
(E) 40

QUESTION 3

Wobble is the relaxation of nucleotide pairing rules of which bases as the codon and anticodon interact?

(A) the third base pair
(B) the first base pair
(C) the second base pair
(D) the first and third base pairs
(E) the second and third base pairs.

QUESTION 4

Prokaryotic cells couple transcription and translation because

(A) they have no nucleolus
(B) all cellular processes occur in the nucleus
(C) mRNA is translated immediately after transcription
(D) they have more free ribosomes
(E) prokaryotes divide more frequently

QUESTION 5

The steps for protein targeting to the ER occur in what order? A) Polypeptide grows through a pore into ER's lumen. B) Ribosome and mRNA disassemble. C) SRP binds to mRNA's signal peptide. D) SRP releases. E) SRP binds to receptor on ER membrane. F) Polypeptide folds. G) Signal peptide cleaved.

(A) C, D, A, E, F, B, G
(B) C, E, G, A, D, F, B
(C) D, A, B, C, G, F, B
(D) C, E, D, A, G, F, B
(E) D, C, A, G, F, B, E

QUESTION 6

GTP fuels which stages of elongation in translation?

(A) peptide bond formation and codon recognition
(B) peptide bond formation and translocation
(C) formation of initiation complex and translocation
(D) formation of initiation complex and peptide bond formation
(E) codon recognition and translocation

QUESTION 7

Protein factors assist which of the following molecules in translation?

(A) rRNA, amino acids, and tRNA
(B) mRNA, tRNA, and ribosomes
(C) rRNA, tRNA, and mRNA
(D) amino acids, mRNA, and tRNA
(E) ribosomes, amino acids, and tRNA

QUESTION 8

Joining tRNA and amino acid requires the loss of how many phosphates from which molecule?

(A) one from GTP
(B) two from GTP
(C) one from ATP
(D) two from ATP
(E) It requires no phosphates.

ANSWERS

1. B
2. C
3. A
4. C
5. D
6. E
7. B
8. D

HEREDITY AND GENETICS
Bacterial Genetics

BOUNTIFUL BACTERIA

NO NUCLEUS, NO PROBLEM

The bacterial genome is a circular, double-stranded DNA molecule tightly coiled to form a **nucleoid**. Outside this chromosome are smaller, circular DNA molecules called **plasmids** that self-replicate or integrate into the bacterial chromosome.

BACTERIAL REPLICATION

Bacterial chromosomes start replicating at an origin, with the daughter strand elongating in both directions, until termination stops the process.

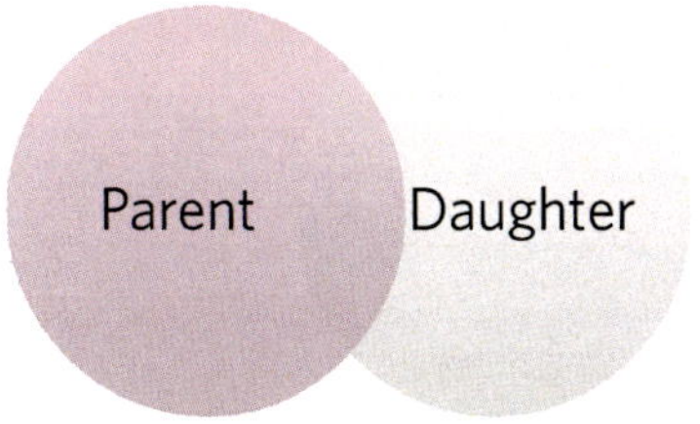

DARE TO BE DIFFERENT

Bacteria reproduce asexually by binary fission, which creates daughter cells that are clones of the parents. Bacteria create genetic diversity by transferring DNA to neighboring cells that integrate the new DNA into their genomes or by shuffling their own genomes.

Process	Mechanism
Transformation	Taking in foreign DNA from the environment and combining it with the bacterial chromosome
Transduction	Problems with prophage excision causes incorporation of some bacterial genes that are transferred to other bacteria
Conjugation	Transfer of genetic material from one bacterium to another
Transposition	Movement of DNA elements from one part of the chromosome to another part

PLASMIDS AND OPERONS

BACTERIAL SEX

During conjugation, two cells form a temporary mating bridge that transfers DNA plasmids between cells.

THE F FACTOR

The F plasmid contains genes necessary for the formation of sex pili. F+ cells (cells with the F plasmid) donate their DNA to F- cells.

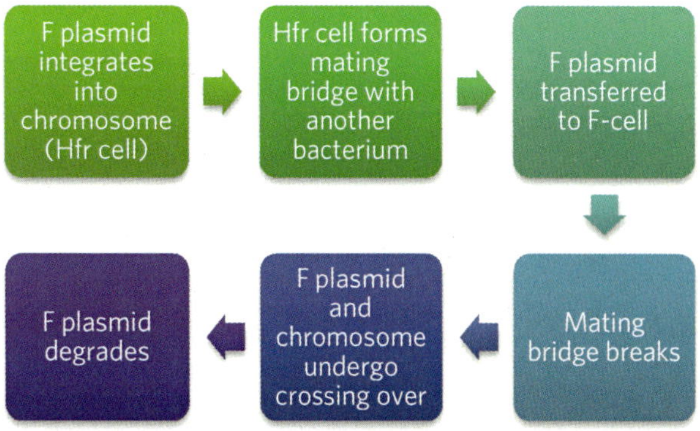

SPOTLIGHT ON TRANSPOSONS

Transposons are complex transposable elements that include several genes flanked by insertion sequences.

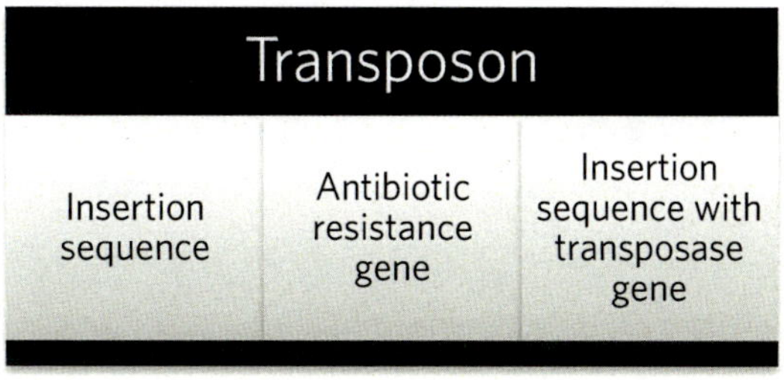

GREAT GROUPING: OPERONS

Bacterial genes are often grouped into units that can be switched off and on. They include an operator (switch), a promoter, and a gene.

- An operator can be turned off by a **repressor** protein that blocks attachment of RNA polymerase to the promoter
- A **corepressor** protein may be needed to activate a repressor
- An **inducer** inactivates a repressor in negative gene regulation
- In positive gene regulation, an **activator** binds upstream of the promoter and facilitates the binding of RNA polymerase to the operon

DEMIDEC®

CRAM QUIZ
Bacterial Genetics

QUESTION 1

During transduction, crossing over between host and viral DNA occurs

(A) when the virus enters the cell
(B) when the virus is assembled
(C) after host DNA is degraded
(D) when the virus reproduces and infects another cell
(E) after prophage formation

QUESTION 2

An Hfr cell has the F plasmid integrated into its chromosome. After conjugation and recombination, the recipient cell becomes

(A) F+
(B) F-
(C) Hfr
(D) a donor
(E) Hfr+

QUESTION 3

What process transfers several antibiotic resistance genes to R plasmids?

(A) transposition
(B) transformation
(C) transduction
(D) insertion
(E) recombination

QUESTION 4

An insertion sequence contains a transposase gene flanked by a series of

(A) four to 10 nucleotides
(B) coding regions that catalyze DNA cutting
(C) inverted repeats
(D) 50 to 250 adenine nucleotides
(E) modified nucleotides

QUESTION 5

Tryptophan regulates its own synthesis by

(A) activating gene expression and inhibiting enzyme activity
(B) inhibiting gene expression
(C) activating gene expression
(D) affecting enzyme activity
(E) inhibiting enzyme activity and gene expression

QUESTION 6

In the presence of an isomer of lactose, the lac operon is

(A) activated
(B) repressed
(C) not induced
(D) not affected
(E) positively regulated

QUESTION 7

When glucose is scare and cAMP in high concentrations, the lac operon

(A) is activated by cAMP
(B) synthesizes mRNA
(C) is repressed
(D) is inhibited by CAP
(E) is negatively regulated

QUESTION 8

Transformation is facilitated by

(A) potassium ions
(B) phage DNA
(C) surface proteins
(D) transposable elements
(E) pores in the membrane

ANSWERS

1. D
2. B
3. A
4. C
5. E
6. A
7. B
8. C

HEREDITY AND GENETICS
Gene Regulation

BEYOND BASE PAIRING

FROM HELICES TO COILS

DNA structure in eukaryotes begins with the double helix and ends with a complex organization of DNA and protein.

LEVELS OF DNA PACKING

DNA packing creates a compact structure that fits large amounts of DNA in small spaces and affects gene expression.

- Nucleosomes
- Chromatin fiber
- Looped domains
- Metaphase chromosome

REGULATING CHROMATIN

DNA's compact form and specific enzymes affect gene replication and transcription.

Histone acetylation	•Creates looser chromatin structure •Transcription takes place
DNA methylation	•Causes gene inactivation
Epigenesis	•Passing of chromatin modifications to next generation
Control elements	•Noncoding DNA regions that regulate transcription
Transcription factors	•Activate, enhance, or repress transcription

TRANSCRIPTS AND PROTEINS

AFTER TRANSCRIPTION

Different RNA molecules play key roles in RNA processing and other regulatory functions.

Transcript
- Changes RNA splicing

miRNA siRNA
- Degrades mRNA
- Prevents translation

PROTEIN PROCESSING

Protein modification helps cells adjust to changes in the environment without affecting transcription. They include...

- Chemical changes, cleavage, and transportation to destination
- Degradation by proteasomes

OUT OF CONTROL

CANCER GENES

Mutations of genes that control cell growth and division can lead to uncontrollable growth and possibly cancer.

Gene	Function
Oncogenes	Cause tumors Found in retroviruses
Proto-oncogenes	Regulate cell growth and division Found in animals
Tumor-suppressor genes	Inhibit cell division
Ras gene	Stimulates cell cycle
P53 gene	Halt cell cycle Stimulate apoptosis

CRAM QUIZ
Gene Regulation

QUESTION 1

Most of the human genome is made up of repetitive DNA sequences. Most of these sequences are

(A) multigene families
(B) introns
(C) regulatory sequences
(D) transposable elements
(E) *alu* elements

QUESTION 2

After the formation of RNA, the movement of retrotransposons is similar to

(A) retrovirus reproduction
(B) transposons
(C) RNA splicing
(D) phage infection
(E) multigene replication

QUESTION 3

The duplication of genes during meiosis caused by unequal crossing over produces

(A) two chromatids with extra genes
(B) one chromatid with two copies of a gene and one chromatid with no copies of a gene
(C) two chromatids with no copies of a gene
(D) different gene sequences
(E) incorporation of transposable elements into chromatids

QUESTION 4

Identical genes of multigene families code for RNA except in the case of

(A) hemoglobin
(B) ribosomal proteins
(C) enzymes
(D) retransposons
(E) histone proteins

QUESTION 5

In what order does protein degradation by proteasomes occur? A) Enzymes cut proteins into peptides. B) Proteasome recognizes ubiquitinated protein. C) Ubiquitin attaches to protein. D) Protein unfolds. E) Peptides degraded in cytosol. F) Protein placed in proteasome's cavity.

(A) F, C, B, E, D, A
(B) F, C, B, A, D, E
(C) C, B, D, F, A, E
(D) C, B, F, D, A, E
(E) C, B, A, F, D, E

QUESTION 6

An miRNA molecule is held together by hydrogen bonds that are cut into fragments, some of which may

(A) reform into miRNA
(B) bind to mRNA
(C) hydrogen bond with any mRNA
(D) synthesize mRNA
(E) bind to DNA

QUESTION 7

In eukaryotes, transcription factors called activators stimulate transcription by causing

(A) DNA bending
(B) other proteins to bind to an enhancer
(C) RNA polymerase to bind to the operator
(D) the unfolding of DNA
(E) stabilization of the DNA molecule

QUESTION 8

Genomic imprinting is caused by

(A) epistasis
(B) epigenic inheritance
(C) histone acetylation
(D) DNA methylation
(E) RNA processing

ANSWERS

1. D
2. A
3. B
4. E
5. C
6. B
7. A
8. D

HEREDITY AND GENETICS
DNA Technology

ENGINEERING THE FUTURE

CLONING IS CRUCIAL

Scientists manipulate genes for different purposes: to understand cellular functions, treat disease, or create a pest-resistant plant. The foundation of these experiments is DNA cloning—the production of many copies of the same gene.

A CLOSER LOOK AT CLONING

DNA cloning uses bacterial plasmids to replicate and transfer genes of interest.

RESTRICTIVE REACTIONS

DNA cloning is possible because of enzymes called **restriction endonucleases.**

1. Enzymes cut both DNA strands at specific locations called **restriction sites** to create **restriction fragments**.
2. The staggered end of the sugar-phosphate backbone (sticky end) base pairs with other sticky ends.
3. DNA ligase permanently binds fragments to other DNA molecules.
4. Recombinant DNA molecules are created.

DID IT WORK?

To double check that a recombinant plasmid functions properly and is taken up by bacteria, scientists insert an antibiotic resistance gene into the plasmid. If all goes well, cells grown in the presence of the antibiotic should develop normally.

A LIBRARY WITHOUT BOOKS

BACTERIAL "BOOKS"

Genomic and cDNA libraries store an organism's genetic material in clones of bacteria or a phage carrying fragments of DNA from the organism.

Genomic Library

- Organism's DNA inserted into plasmids
- Clones with plasmids grown
- DNA spans entire genome

cDNA Library

- mRNA extracted from cell
- Reverse transcriptase produces DNA
- Restriction enzymes modify cDNA
- cDNA inserted into plasmid

TERRIFIC TECHNIQUES

Scientists use several techniques in genetic engineering to study DNA molecules.

PCR: POLYMERASE CHAIN REACTION

PCR is a quick method for replicating large quantities of poor samples (scarce or impure) of DNA.

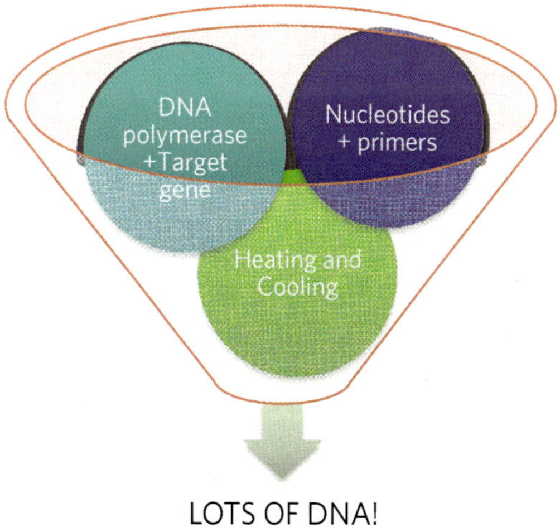

GEL ELECTROPHORESIS

This technique separates DNA molecules that differ in size.

1. A current is passed through the gel.
2. Small molecules travel faster than larger molecules down the gel.
3. A DNA-binding dye illuminates the bands on the gel.

CRAM QUIZ
DNA Technology

QUESTION 1

Gene therapy is the
(A) creation of an individual's DNA markers
(B) diagnosis of disease by genetic marker
(C) injection of clone genes into a patient
(D) production of large quantities of protein
(E) removal of wastes with engineered bacteria

QUESTION 2

Why is the T1 plasmid useful for the genetic engineering of plants?
(A) It does not degrade readily.
(B) It can be cut by several different restriction enzymes.
(C) It contains many antibiotic resistance genes.
(D) It contains the gene that codes for transposase.
(E) It is easily incorporated into the plant's genome.

QUESTION 3

To determine gene function, scientists perform
(A) RNA interference
(B) *in vivo* mutagenesis
(C) DNA sequencing
(D) DNA mapping
(E) restriction modification

QUESTION 4

Yeast artificial chromosomes (YACs) are used in eukaryotic cloning because
(A) they contain transposable elements
(B) it contains the important elements of a eukaryotic chromosome
(C) it has an origin of replication, a centrosome, and two telomeres
(D) it is similar to phage plasmids
(E) they already contain foreign DNA

QUESTION 5

Electroporation introduces foreign DNA into cells by
(A) changing the voltage across the plasma membrane
(B) causing the release of calcium ions from the ER
(C) sending an electrical pulse through the cell
(D) opening pores in the plasma membrane
(E) injection DNA into the cells

QUESTION 6

How do probes identify genes of interest?
(A) with a radioactive isotope
(B) with a special nucleotide sequence
(C) by cutting the gene from the plasmid
(D) by covalently bonding to the gene
(E) by binding to DNA and RNA that are not of interest

QUESTION 7

Differences in which of the following cause different restriction fragment length polymorphisms (RFLPs)?
(A) crossing over of sister chromatids
(B) exon excisions
(C) restriction sites of homologues
(D) RNA splicing
(E) DNA condensation

QUESTION 8

Put the stages of genetic mapping in order: A) Ordering of large overlapping fragments. B) Genetic marker ordering. C) Sequencing of fragments. D) Ordering of small overlapping fragments. E) Sequence assembly.
(A) B, D, A, E, C
(B) A, D, B, C, E
(C) B, D, A, C, E
(D) A, C, E, B, D
(E) B, A, D, C, E

ANSWERS

1. C
2. E
3. A
4. B
5. D
6. A
7. C
8. E

HEREDITY AND GENETICS
Evolution

REVOLUTIONARY DESCENT

CHANGING HISTORY

The publication of Darwin's *On the Origin of Species by Means of Natural Selection* tore down older theories explaining natural history and biological diversity.

OUT WITH THE OLD

Aristotle's beliefs that species never changed greatly influenced later evolutionary theories until Darwin's time.

Aristotle	•Scala naturae's linear heirarchy
Linnaeus	•Taxonomy's grouping similar species
Cuvier	•Catastrophism kills many species at once
Hutton	•Gradualism's cumulative, slow processes
Lyell	•Uniformitarianism's conserved geologic processes
Lamarck	•Used traits grow stronger and are passed on

VOYAGE OF A LIFETIME

Darwin's trip to South America and the Galapagos Islands sparked several observations and conclusions about species diversity that led to the development of the theory of evolution.

•Populations could grow infinitely
•But they don't
•Because resources are limited
→ Limited resources affects survival

•No two individuals are alike
•Variation is passed on
→ Survival depends on fitness and inherited trais

Conclusion → Different levels of fitness and traits change a population's makeup over time.

BE SELECTIVE

EVOLUTION IN A NUTSHELL

- ✓ Individuals with certain traits outcompete others for resources (NATURAL SELECTION)
- ✓ They reproduce, passing the traits on to their offspring
- ✓ The environment changes, and some adapt to the new conditions (ADAPTATION)
- ✓ Over time, this process alters populations and may create new species.(DESCENT WITH MODIFICATION)

EVOLUTION IN ACTION

Scientists study the evolutionary process by studying living populations and the fossil record.

Homology	▪ Species with similar traits from shared ancestry ▪ Anatomical or molecular ▪ Example: vestigial organs
Biogeography	▪ Related species populate a region ▪ Endemic species are found in one place ▪ Example: island species
Fossil record	▪ Supports shared ancestry and descent from a common organism ▪ Example: aquatic mammals' ancestors lived on land

THE YOUNG EARTH

THE EVOLUTION OF SIMPLE CELLS

1. Electron-adding (reducing) atmosphere filled with water and inorganic compounds creates small organic molecules.
2. Amino acids combine and form polymers.
3. Formation of protobionts (molecules surrounded by a membrane with a different internal environment).
4. RNA molecules (ribozymes) self-replicate.
5. Early prokaryotes appear as protobionts, metabolize molecules, and harness light energy (heterotrophs).
6. Heterotrophs increase atmospheric oxygen levels.

CRAM QUIZ
Evolution

QUESTION 1

Which of the following applies to the theory of evolution?

(A) Individuals and populations evolve.
(B) Acquired traits that are favorable are passed on to offspring.
(C) The favorability of traits varies with environmental conditions.
(D) About 70% of species are extinct.
(E) The theory doesn't take artificial selection into consideration.

QUESTION 2

The field of paleontology was founded in conjunction with the theory of

(A) catastrophism
(B) gradualism
(C) use and disuse
(D) uniformitarianism
(E) taxonomy

QUESTION 3

Vestigial organs are

(A) examples of analogous structures
(B) rarely found in animals
(C) dissimilar structures found in closely related organisms
(D) ancient structures
(E) only detectable at the molecular level

QUESTION 4

Why is evolution a theory and not a law?

(A) It is a hypothesis.
(B) It cannot be disproven.
(C) It requires more testing before it can become a law.
(D) Some parts of the original theory have been disproven.
(E) Scientists are skeptical in order to prevent any theory from becoming a doctrine.

QUESTION 5

Evidence supporting evolution demonstrates that natural selection is

(A) a creative process
(B) based in a time and place
(C) apparent only over a long period of time
(D) a process that cannot be tested, only observed
(E) not a unifying process

QUESTION 6

How do pseudogenes preserve genetic variation?

(A) by changing frequently due to natural selection
(B) by not affecting an organism's cellular processes
(C) by not being affected by natural selection
(D) by never affecting survival
(E) by becoming activated

QUESTION 7

Changes in allele frequencies in a population over time are due to

(A) mutations
(B) gene duplication
(C) sexual recombination
(D) gene flow
(E) random fertilization

QUESTION 8

Which of the following phenomena is most predictable?

(A) natural selection
(B) bottleneck effect
(C) founder effect
(D) genetic drift
(E) gene flow

ANSWERS

1. C
2. A
3. D
4. E
5. B
6. C
7. D
8. A

HEREDITY AND GENETICS
Mechanisms of Evolution

STARTING SMALL

EVERYONE IN THE POOL!

Microevolution follows the change in a population over time. This group's collection of genes (**gene pool**) varies with environmental conditions. Some alleles may be homozygous (the same) or heterozygous (two or more) for a trait.

TWO SCIENTISTS, ONE THEORY

The **Hardy-Weinberg theorem** provides a point of reference against which evolving gene pools are measured. It states that

When alleles segregate independently, their frequencies stay the same from one generation to the next, protecting genetic variation.

LEARN YOUR FORMULAS!

How do you know if a population is in a state of Hardy-Weinberg equilibrium?

Double check this with the equation:

$$p^2 + 2pq + q^2 = 1,$$

where p and q represent allele frequencies.

SMALL CHANGES

Gene pools change by two methods: sexual recombination and mutation.

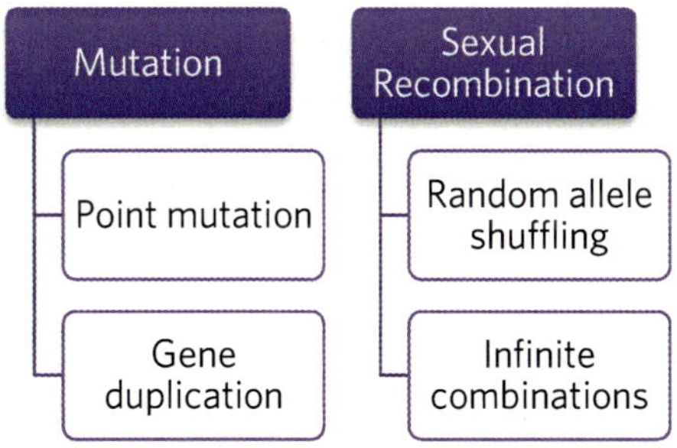

BIG CHANGES

Three processes affect allele frequencies and cause evolutionary changes.

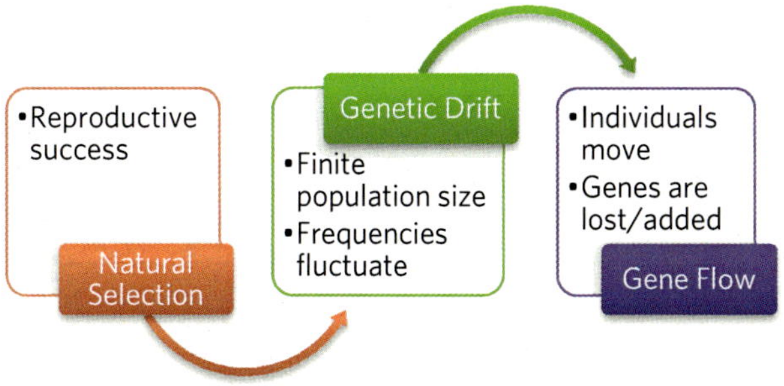

THE NATURAL WAY

A CLOSER LOOK: NATURAL SELECTION

Natural selection is the most powerful and important process in evolution, but it depends on genetic differences among individuals.

KEY CONCEPTS

Genetic variation is based on differences in characters (**polymorphisms**) and gene pools (**geographic variation**).

- **Fitness** is a measure of an individual's reproductive success
- **Relative fitness** measures a genotype's contribution to the gene pool compared to other genotypes

TYPES OF SELECTION

Phenotypes in a population occur along a spectrum, with some traits being common and others rare. Natural selection is placed in three different categories based on whether it favors common or rare phenotypes.

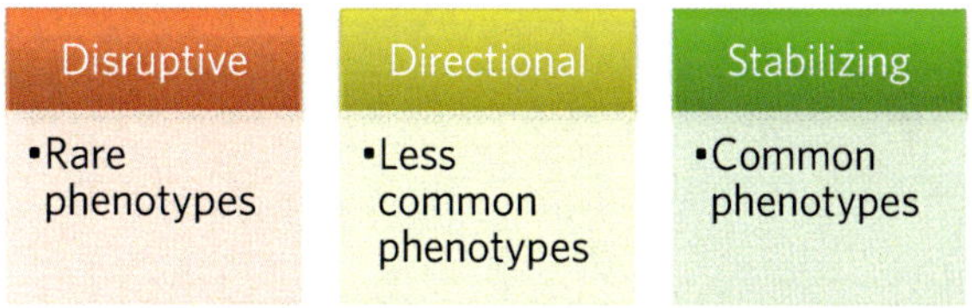

TUG OF WAR

The question: Why doesn't natural selection reduce variation by getting rid of unfavorable genes?

The answer: Several factors preserve variation.

Factor	*Function*
Frequency-dependent selection	Trait less favorable as it becomes more common
Heterozygous advantage	Heterozygotes favored over homozygotes, preserving multiple alleles
Diploidy	Preserves recessive alleles
Neutral variation	Alleles not affected by natural selection

CRAM QUIZ
Mechanisms of Evolution

QUESTION 1

The Hard-Weinberg theorem examines a population that is

(A) very diverse
(B) not evolving
(C) evolving quickly
(D) evolving slowly
(E) small

QUESTION 2

In the Hardy-Weinberg equation, q^2 represents the

(A) frequency of heterozygotes
(B) frequency of all alleles
(C) frequency of male gametes
(D) frequency of female gametes
(E) frequency of homozygotes

QUESTION 3

A population in a state of Hardy-Weinberg equilibrium must not have

(A) genetic drift, random mating, artificial selection, or large population size
(B) recombination, mutation, or gene flow
(C) natural selection, artificial selection, or small population size
(D) gene flow, mutations, natural selection, or small population size
(E) random mating, gene flow, mutations, or small population size

QUESTION 4

Sexual selection is based on

(A) mating choices and competition for mates
(B) sexual dimorphisms and competition
(C) reproductive success and sexual competition
(D) mating success and sexual dimorphisms
(E) choosing fitter mates and reproductive success

QUESTION 5

If a population's gene pool has 23,070 loci and its average heterozygosity is 23%, how many loci are homozygous?

(A) 5,306
(B) 2,653
(C) 17, 764
(D) 8,882
(E) 23,070

QUESTION 6

What may be the advantage of sexual selection?

(A) It allows for subsequent generations' adaptation.
(B) It may benefit organisms in the short term.
(C) It out-produces asexual reproduction.
(D) Females choose the best mate.
(E) It provides more adaptations over time.

QUESTION 7

What is the difference between the bottleneck and founder effects?

(A) Founder effect occurs when a population is founded, and bottleneck effect occurs when a new population begins to grow.
(B) The found effect depends on environmental conditions, and the bottleneck effect depends on population size.
(C) The founder effect refers to small populations, and the bottleneck effect refers to growing populations.
(D) The bottleneck effect occurs after an environmental catastrophe, and the founder's effect occurs when a few individuals found a population.
(E) The bottleneck effect is an example of gene flow, and the found effect is an example of genetic drift.

ANSWERS

1. B
2. E
3. D
4. A
5. C
6. B
7. D

HEREDITY AND GENETICS
Evolutionary Patterns

RAISING THE BAR

BEYOND POPULATIONS

Macroevolution describes evolutionary changes that take place at higher taxonomic levels, such as the appearance of opposable thumbs.

IMPORTANT PATTERNS

Evolutionary patterns differ based on the method of speciation.

Anagenesis	•Changes accumulate •Species gradually transforms
Cladogenesis	•Gene pool separates •New species appears

Note: Cladogenesis promotes diversity by creating new species.

MAKE YOUR OWN SPECIES

Speciation is a crucial part of evolution. In general, individuals of the same species can mate and produce viable, fertile offspring.

1. Create a barrier to prevent individuals from mating.
2. Choose your barrier.

- Prezygotic
  - Habitat
  - Time of day or year
  - Behavior
  - Mechanics
  - Gametes
- Postzygotic
  - Offspring not viable
  - Offspring not fertile
  - Second generation not viable or fertile

3. Decide how to reduce gene flow.

Allopatric	Separate species geographically	
Sympatric	Alter chromosomes/ mating	Change resources

Congratulations! You created many new species from one population. You win the **Adaptive Radiation** prize!

TIMING IS EVERYTHING

DISAGREEING WITH DARWIN

Darwin theorized that species appear gradually over time, but not all scientists agree.

PROBLEMS WITH GRADUALISM

- Gradualism doesn't explain why species appear suddenly in the fossil record, remain unchanged, and then vanish
- **Punctualism** states that species change the most when they first appear

PROBLEMS WITH PUNCTUALISM

1. The fossil record may be incomplete.
2. The fossil record can't show what happens over shorter time periods.
3. The fossil record only shows anatomical changes.

SMALL GENES, BIG CHANGES

Even small changes to the genes that control development cause big morphological changes in its timing, pacing, and patterning.

- Heterochrony
  - Changes rate or timing of development
    - Allometric growth
      - Growth rates of different body parts
        - Chimp vs. human skull
    - Paedomorphosis
      - Timing of sexual maturity
        - Retain juvenile features
- Body Patterning
  - Changes orientation or placement of body parts
    - Homeotic genes
      - Determine body patterning
        - *Hox* genes and limb formation

CRAM QUIZ
Evolutionary Patterns

QUESTION 1

How do the *Hox* genes relate to the evolution of vertebrate animals?

(A) They developed from unrelated genes found in invertebrates.
(B) They appeared spontaneously at the same time as vertebrates.
(C) A deletion mutation altered their sequence in invertebrates.
(D) They are duplicated twice in vertebrates.
(E) They provided anatomical complexity.

QUESTION 2

How does autopolyploidy cause reproductive isolation in one generation?

(A) Sterile hybrids can't become fertile.
(B) Plants of different species can't produce offspring.
(C) Polyploids can self-pollinate, but they can't mate with diploid individuals.
(D) Polyploids contain mutations that prevent it from reproducing.
(E) Additional chromosomes don't affect reproductivity.

QUESTION 3

The inability of second-generation hybrid species to persist is due to

(A) an accumulation of recessive alleles
(B) reduced fertility
(C) prezygotic barriers
(D) mating with individuals of their own species
(E) hampered fertilization

QUESTION 4

Which of the following developmental genes is a segmentation gene in *Drosophila*?

(A) egg-polarity
(B) pair-rule
(C) homeotic
(D) *Hox*
(E) maternal effect

QUESTION 5

What is the function of homeotic genes?

(A) differentiation
(B) segment formation
(C) axis formation
(D) limb formation
(E) initiation of metamorphosis

QUESTION 6

What switches genes off and on during plant and animal development?

(A) cell signal molecules
(B) promoters
(C) transcription factors
(D) second messengers
(E) primers

QUESTION 7

The homeobox codes for a protein that performs what function?

(A) alters proteins
(B) binds to DNA
(C) regulates transcription
(D) catalyzes reactions
(E) part of the plasma membrane

QUESTION 8

Development gene are conserved in many species because they

(A) appeared early in evolutionary history
(B) mutated many times
(C) are molecular analogies
(D) perform necessary functions
(E) are switched on in the embryonic stage

ANSWERS

1. E
2. C
3. A
4. B
5. D
6. C
7. B
8. A

POPULATIONS
Phylogeny and Systematics

THE TREE OF LIFE

STRANGE SIBLINGS

Since Aristotle, scientists have tried to reconstruct natural history by classifying organisms based on similarities.

COMMON GROUND

Phylogeny traces a species' natural history back to a common ancestor by examining morphological, molecular, and fossil evidence.

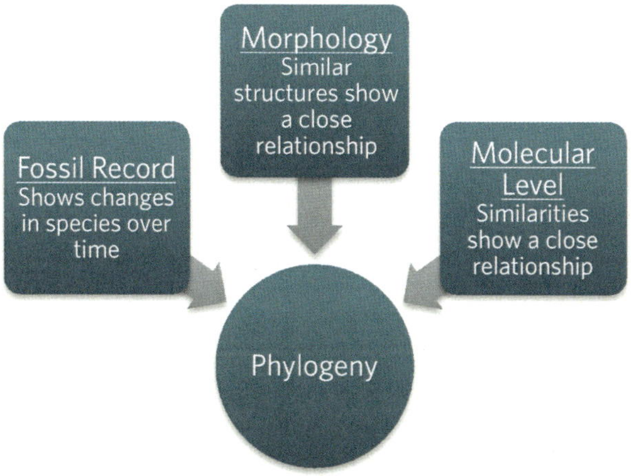

CURTAILING CONFUSION

Although similarities among species (homologies) reflect common ancestry, they can be confused with analogies (similar adaptations not based on ancestry).

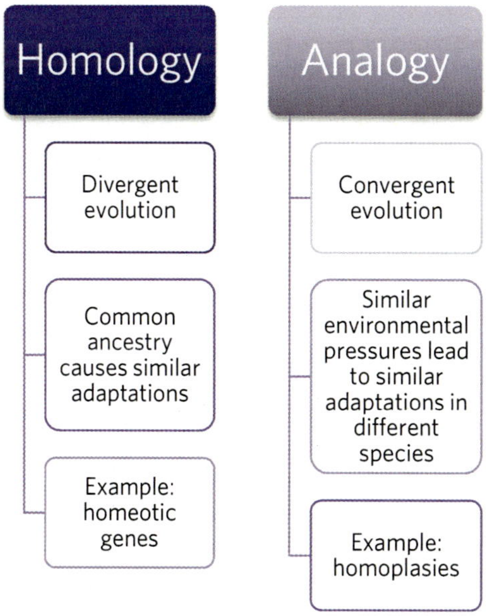

CRAZY FOR CLASSIFICAITON

SYSTEMS OF SIMILARITIES

Classification systems group species together by focusing on different shared characteristics.

CLASSIC HIERARCHY

Linnaeus grouped species into eight progressively broader categories, but this method is also used to create phylogenic trees.

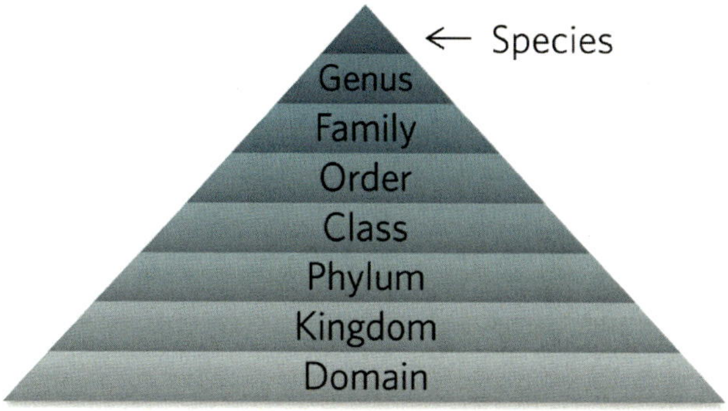

CLADISTICS

Another system of classification groups species based on homologies, with each clade (branch of the tree) representing common ancestors.

HOW TO CREATE A CLADOGRAM

1. Decide if the group is a clade (monophyletic).
2. If not, are some descendants included (paraphyletic) or is there no common ancestry (polyphyletic)?
3. Examine the homology.
4. Is it a trait that's shared with many species outside the group (shared primitive character), or is it unique to the group (shared derived character)?
5. Differentiate the characters.
6. Designate a group similar to the one being studied (outgroup and ingroup, respectively).
7. Compare groups to create cladogram.

TIMING OF SPECIATION

Phylograms and ultrametric trees compare DNA sequences among species.

PHYLOGRAMS
Different branch lengths reflect rates of evolution.

ULTRAMETRIC TREES
Evolutionary changes placed in context of fossil record.

CRAM QUIZ
Phylogeny and Systematics

QUESTION 1

When classifying many different species, it is important to create a hypothesis that assumes the fewest molecular changes. This approach is called

(A) maximum likelihood
(B) neutral theory
(C) maximum parsimony
(D) taxonomy
(E) systematics

QUESTION 2

Aligning DNA sequences to establish molecule homologies may require

(A) removing insertions or deletions
(B) inserting homologous sequences
(C) removing dissimilar sequences
(D) inserting inverted repeats
(E) filling in gaps

QUESTION 3

Gene duplication is important to evolution because it

(A) creates genetic diversity
(B) occurs frequently
(C) is carried out by transposons
(D) increases the number of genes
(E) occurs spontaneously

QUESTION 4

What is the difference between orthologous and paralogous genes?

(A) Paralogous genes are passed from one generation to the next.
(B) Paralogous genes are caused by duplication.
(C) Orthologous genes are found in multiple places throughout the genome.
(D) Orthologous genes are caused by duplication.
(E) Paralogous genes are the result of speciation.

QUESTION 5

Neutral theory assumes that

(A) molecular clocks run at a constant rate
(B) DNA evolves at a constant rate
(C) most evolutionary change is the result of genetic drift
(D) changes to genes and proteins rarely affect survival
(E) less important genes evolve more quickly

QUESTION 6

Molecular clocks and neutral theory do not account for

(A) the timing of evolution
(B) changes in DNA from gene flow
(C) the rate of evolution
(D) the relationship between DNA and evolution
(E) changes in DNA due to natural selection

QUESTION 7

Genetic sequences are used to decipher phylogeny when

(A) horizontal gene transfer occurs
(B) they are part of a genome
(C) homologies can't be identified
(D) they evolved slowly
(E) the sequences align

QUESTION 8

According to rRNA sequencing and comparison across species, organisms are grouped

(A) into three domains
(B) into five kingdoms
(C) based on the assumption that genes move horizontally
(D) based on maximum parsimony
(E) hypothetically

ANSWERS

1. C
2. A
3. D
4. B
5. D
6. E
7. D
8. A

POPULATIONS
Prokaryotes

CREATURES GREAT AND SMALL

SUPERIOR REVISIONS

In recent years, DNA sequencing and other molecular data have caused scientists to rethink the classification of certain species and create a **three-domain system** above kingdoms (superkingdoms).

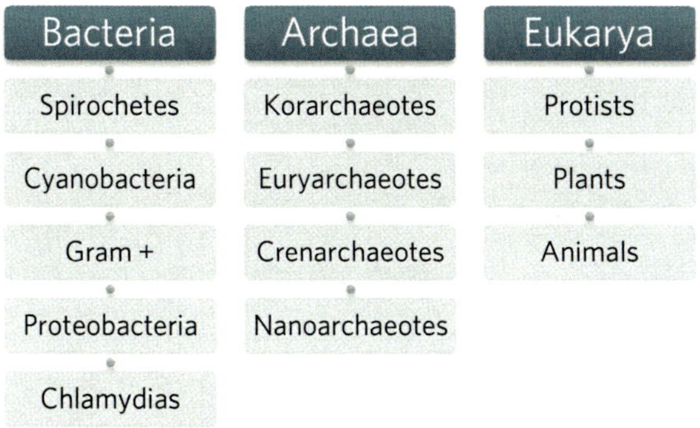

PLENTIFUL PROKARYOTES

Prokaryotes make up the two superkingdoms Bacteria and Archaea, and have several defining characteristics:

- Unicellular
- Small
- Spherical (cocci), rod (bacilli), and spiral-shaped

A CLOSER LOOK

Prokaryotes differ from other organisms in cell surface, genome, and internal cell structuring.

1. A polysaccharide or protein coating (**capsule**) may surround cell walls (with or without appendages) to facilitate attachment to host cells or other prokaryotes.

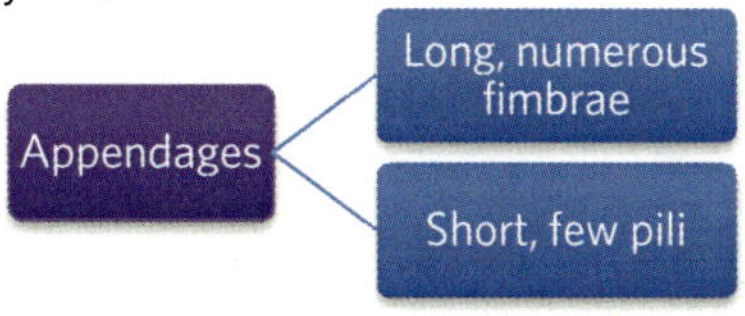

2. Prokaryotes have no internal compartments but may have specialized membranes for metabolism.

Prokaryotic DNA is organized differently than eukaryotes'.

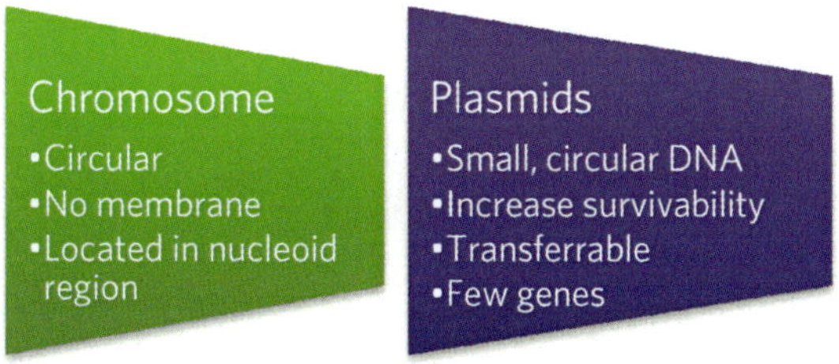

BRED FOR SUCCESS

WORLD DOMINATION

Prokaryotes live nearly everywhere, a feat due to their successful adaptation to different environments.

FAST MOVES

Half of all prokaryotes move toward or away from stimuli (***taxis***), usually with the help of flagella.

Flagella Structure

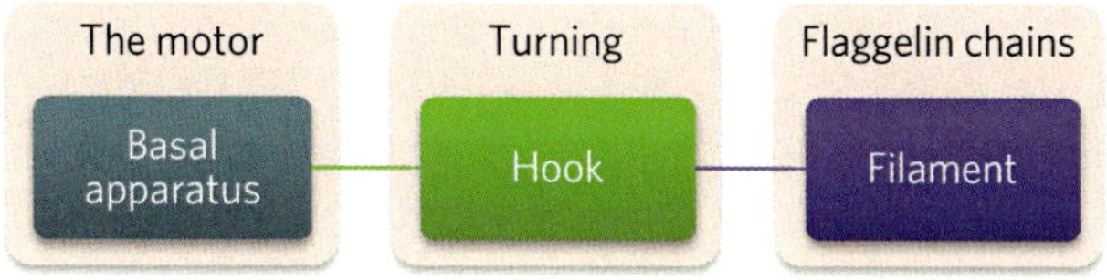

QUICK CHANGES

Prokaryotic populations adapt quickly to the environment.

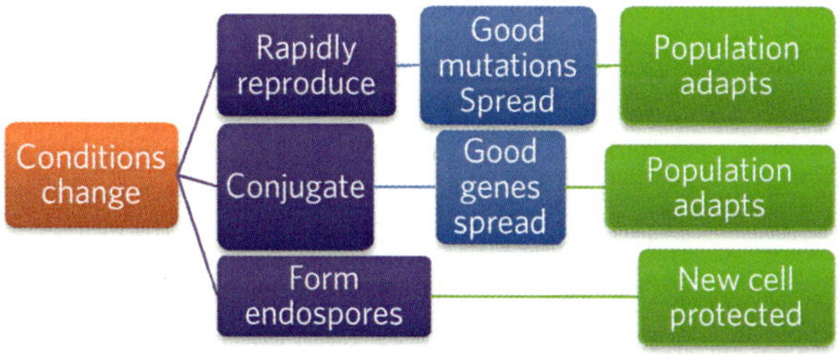

MANY METABOLISMS

Prokaryotes use a variety of nutrients to fuel metabolism; they differ in their energy and carbon sources and in their uses of oxygen and nitrogen.

Organism	Energy	Carbon source
Photoautotroph	Light	CO_2
Chemoautotroph	Inorganic compounds	CO_2
Photoheterotroph	Light	Organic compounds
Chemoheterotroph	Organic compounds	Organic compounds

ARCHAEA

Archaea are ancient organisms that often live in extreme environments (**extremophiles**).

- Extreme **thermophiles** thrive in hot climes
- Extreme **halophiles** thrive in high salinity

CRAM QUIZ
Prokaryotes

QUESTION 1

Prokaryotic and eukaryotic flagella differ in that

(A) eukaryotic flagella are narrow
(B) prokaryotic flagella are not covered by a plasma membrane
(C) flagella movement is powered by ATP only in prokaryotes
(D) eukaryotic flagella are composed of globular proteins
(E) prokaryotic flagella have a wavelike motion

QUESTION 2

Antibiotics kill bacteria without harming eukaryotes because

(A) they penetrate prokaryotes' capsules
(B) eukaryotic membranes have no receptors to bind antibiotics
(C) prokaryotic DNA has no histones
(D) they bind to prokaryotic ribosomes
(E) they block prokaryotic transcription

QUESTION 3

Endospores cannot be killed by boiling water because they are surrounded by a touch wall that forms after

(A) the prokaryote reproduces
(B) water is removed from the prokaryote
(C) the prokaryote's metabolism slows
(D) the original prokaryote dies
(E) the prokaryote's chromosome duplicates

QUESTION 4

Obligate anaerobes use fermentation and anaerobic respiration for metabolism. What ions replace oxygen as electron acceptors in anaerobic respiration?

(A) nitrate ions
(B) potassium ions
(C) cytochromes
(D) sulfhydryl ions
(E) ADP

QUESTION 5

Which of the following resources do nitrogen-fixing cyanobacteria need to survive?

(A) oxygen, nitrogen, light
(B) carbon dioxide, nitrogen, light
(C) nitrogen, water, carbon dioxide, light
(D) organic compounds, nitrogen, light, water
(E) water, nitrogen, oxygen, light

QUESTION 6

Members of which domain(s) has DNA with histones?

(A) Bacteria and Eukarya
(B) Archaea and Eukarya
(C) Eukarya
(D) Bacteria and Archaea
(E) Bacteria

QUESTION 7

Introns are present in the DNA sequences of members of which domain(s)?

(A) Bacteria
(B) Bacteria and Archaea
(C) Eukarya
(D) Archaea and Eukarya
(E) Bacteria and Eukarya

QUESTION 8

Different forms of RNA polymerase are found in the members of which domain(s)?

(A) Eukarya
(B) Bacteria and Eukarya
(C) Eukarya and Archaea
(D) Bacteria and Archaea
(E) Bacteria

ANSWERS

1. B
2. D
3. E
4. A
5. C
6. B
7. D
8. C

POPULATIONS
Protists

WATERY WORLDS

FALLEN KINGS

For centuries, microscopic creatures called protists have intrigued scientists. Once part of their own kingdom, protists are now classified with eukaryotes thanks to advances in molecular biology.

THE TINIEST EUKARYOTES

Some protists are more closely related to animals or plants than other protists, but, in general, they have several defining characteristics.

- Unicellular; some are multicellular or colonial
- Complex
- Diverse metabolisms (**mixotrophs** use photosynthesis and respiration)
- Mostly aquatic
- Diverse methods of reproduction

MAJOR GROUPS

Protists are segregated into groups based on unique structures or ancestry.

Group	*Unique structure*
Diplomonads/ Parabasalids	Mitochondria
Cerozoans/ Radiolarians	Pseudopodia, tests
Amoebozoans	Pseudopodia
Stramenopiles	Flagella
Alveolates	Alveoli
Euglenozoans	Kinetoplasts, flagella
Red/green algae	Relative of plants

DIPLOMONADS AND PARABASALIDS

Mostly **anaerobic** protists have no plastids and unique mitochondria.

Unique Mitochondria				
No DNA	No electron transport chain	No enzymes typical to citric acid cycle	May be smaller in size	May make cofactors needed for ATP production

SMALL BUT COMPLEX

CERCOZOANS AND RADIOLARIANS/AMEOBOZOANS

1. Protists have threadlike or lobe-shaped **pseudopodia**—cellular extensions responsible for feeding and movement.
2. Forams and radiolarians have calcium carbonite or silica outer shells called **tests**.
3. Slime molds form a mass of cytoplasm with many nuclei called **plasmodium**.

EUGLENOZOANS

Parasites, predators, and autotrophs have unique flagella and mitochondria.

Mitochondria
- Disk-shaped cristae
- Mass of DNA (kinetoplastid)

Flagella
- Spiral or crystalline rod inside

ALVEOLATES

Protists have a sac located beneath the plasma membrane. Ciliates have micronuclei for conjugation and cilia for movement and feeding. Apicomplexans have organelles for invading hosts (apical complexes).

STRAMENOPHILES/RED AND GREEN ALGAE

Protists have two flagella—one with projections and the other smooth—paired together. They include brown algae (seaweed), green algae (plankton), oomycetes (water mold), and diatoms. Red algae are the largest seaweeds. Green algae may live symbiotically as **lichens**.

WATER MOLD'S LIFE CYCLE

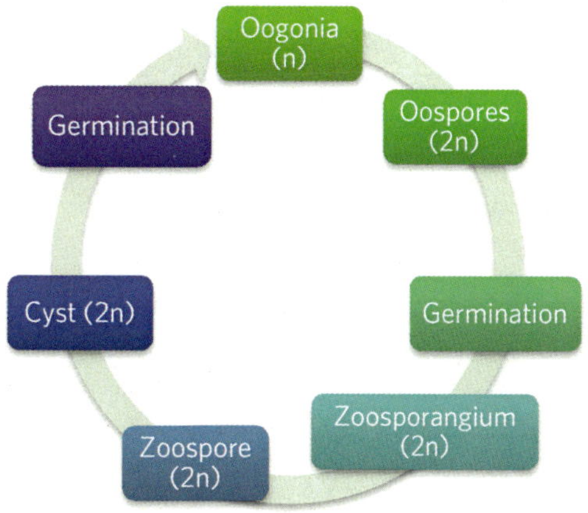

CRAM QUIZ

Protists

QUESTION 1

Ciliates reproduce by binary fission and conjugation. Which step follows micronuclear fusion?

(A) the macronucleus disintegrates
(B) the micronuclei undergo four mitotic divisions
(C) binary fission
(D) meiosis
(E) cell separation

QUESTION 2

Oomycetes differ from fungi in that

(A) the diploid state is predominant in fungi
(B) fungi have cilia
(C) oomycetes' cell walls are made of cellulose
(D) they evolved divergently
(E) oomycetes have plastids

QUESTION 3

What similarities do diatoms, green algae, and brown algae share?

(A) They store food as oil.
(B) They store food as glucose.
(C) They store food as glycogen.
(D) They store food as protein.
(E) They store food as laminarin.

QUESTION 4

How do heteromorphic and isomorphic life cycles differ?

(A) Eggs are released in heteromorphic life cycles.
(B) In heteromorphic life cycles, the structures of sporphytes and gametophytes differ.
(C) In isomorphic life cycles, sporphytes and gametophytes have the same chromosome number.
(D) Syngamy occurs after fertilization in isomorphic life cycles.
(E) Syngamy does not occur in heteromorphic life cycles.

QUESTION 5

How do the life cycles of plasmodial and cellular slime molds differ?

(A) The feeding stage of plasmodial slimes molds is haploid.
(B) Plasmodial slime molds aggregate to form fruiting bodies.
(C) Cellular slime molds produce flagellated, haploid cells.
(D) Cellular slime molds undergo asexual reproduction.
(E) Cellular slime molds produce spores.

QUESTION 6

Red algae contain phycoerythin, a pigment that penetrates water at great depths. What colors of light does this pigment absorb?

(A) blue and green light
(B) red and blue light
(C) green and blue light
(D) violet and red light
(E) violet and blue light

QUESTION 7

The seaweed body called a thallus contains

(A) plant structures
(B) stems but no roots
(C) plant-like structures
(D) leaves but no roots or stems
(E) roots with no stems

QUESTION 8

The apicomplexan *Plasmodium* causes malaria and spreads through what host by means of sporozoites?

(A) human
(B) mosquito
(C) insects
(D) animals
(E) It spreads by merozoites, not sporozoites.

ANSWERS

1. A
2. C
3. E
4. B
5. D
6. A
7. C
8. B

POPULATIONS
Fungi

MAGNIFICENT MUSHROOMS

FEEDING FRENZY

Fungi eat anything, from paint to plants. They break down organic materials and supply ecosystems with essential inorganic compounds like nitrogen and carbon.

NO STOMACH? NO PROBLEM

Fungi don't consume their food. They secrete **exoenzymes** that hydrolyze organic compounds and absorb the products.

STRANGE STRUCTURES

Multicellular fungi contain filaments (**hyphae**) with cell walls made of chitin interwoven into networks (**mycelium**) that spread through the soil for better absorption. Walls (**septa**) with **pores** separate cells and allow nutrients and organelles to flow freely.

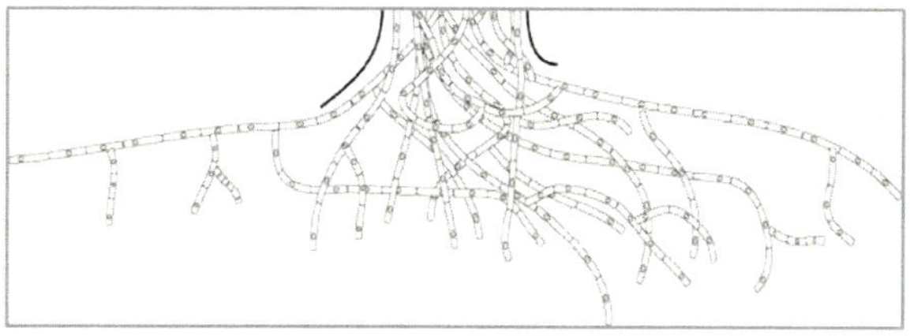

FUNGAL SEX

Many fungi undergo both sexual and asexual reproduction with unique characteristics. The cytoplasm of mycelium fuses first (**plasmogamy**), followed by the nuclei (**karyogamy**).

SEXUAL REPRODUCTION

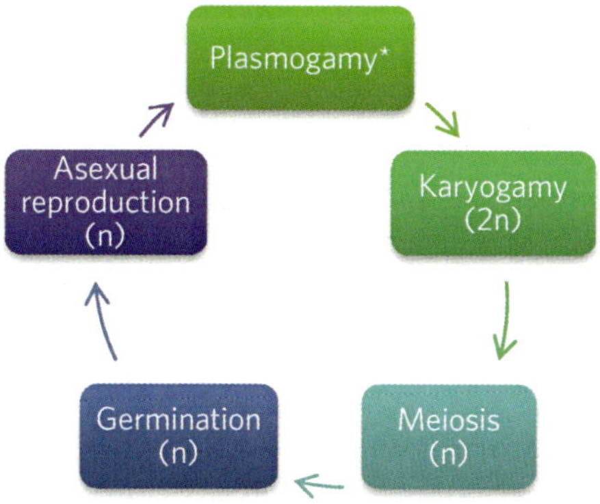

Heterokaryons have many genetically different nuclei.

ASEXUAL REPRODUCTION

THE PHYLOGENY OF FUNGI

THE FIRST FUNGUS

Scientists believe ancestral fungi were unicellular flagellates that moved to land with plants and other organisms.

FUNGAL LINEAGES

Fungi belong to five phyla that differ in body structures and methods of reproduction.

QUICK REFERENCE

Phyla	*Characteristic*
Basidiomycota	Basiciocarp
Ascomycota	Asci
Glomeromycota	Arbuscular mycorrhizae
Zygomycota	Resistant zycosporangium
Chytridiomycota	Flagellated spores

SPOTLIGHT ON SYMBIOSIS

The first terrestrial plants and most plants alive today have a symbiotic relationship with fungi.

Inside plants' roots, fungal hyphae (**mycorrihizae**)...

- Absorb food molecules
- Donate water and nutrients to the plants
- Secrete antibiotics that protect against bacteria
- Secrete growth factors that stimulate root growth

MYCORRIHAE TAKE TWO FORMS

Ectomycorrihae	Endomycorrhize
Mantle encases roots	No mantle
Hyphae grow into soil	Hyphae grow into soil
Do not damage roots	Digest parts of root cells
Hyphae grow into root cells' extracellular spaces	"Infect" root cells
No arbuscules	Form arbuscules
Thicker, shorter, more branching	Shorter, narrower, less branching

CRAM QUIZ

Fungi

QUESTION 1

30% of fungi are parasitic and cause 80% of plant diseases. How does mycosis (fungal infection) damage host cells?

(A) by damaging plasma membranes
(B) by absorbing nutrients
(C) by lysing the cells
(D) by taking over cellular processes
(E) by binding to receptors on the plasma membrane

QUESTION 2

Which of the following structures is contained in lichen composed of ascomycetes and unicellular green algae?

(A) hyphae, algae, and ascocarps
(B) hyphae, soredia, and algae
(C) hyphae, soredia, algae, and asci
(D) hyphae, soredia, algae, and ascocarps
(E) hyphae, algae, ascocarps, and conidia

QUESTION 3

What happens before basciocarp formation in the life cycle of basidiomycetes?

(A) karyogamy
(B) meiosis
(C) plasmogamy
(D) germination
(E) dispersal

QUESTION 4

What happens after the formation of asci in the life cycle of ascomycetes?

(A) karyogamy
(B) meiosis
(C) plasmogamy
(D) germination
(E) dispersal

QUESTION 5

Fungi that form branched structures called arbuscules that penetrate root cells are

(A) chytrids
(B) glymerocetes
(C) zygomycetes
(D) ascomycetes
(E) basidiomycetes

QUESTION 6

How does asexual reproduction in yeasts and molds differ?

(A) Yeasts reproduce by budding.
(B) Molds form heterokaryons.
(C) Yeasts produce spores.
(D) Molds produce conidiophores.
(E) Yeasts form zygosporangium.

QUESTION 7

Sexual reproduction occurs in fungi mycelia of different mating strains. What chemicals do mycelia release that bind to a mate's cell surface?

(A) ethylene
(B) growth factor
(C) pheromones
(D) cytokinins
(E) G proteins

QUESTION 8

What structures allow fungi to penetrate the cell wall or plasma membrane of host cells?

(A) hyphae
(B) mycorrhizae
(C) arbuscules
(D) haustoria
(E) soredia

ANSWERS

1. B
2. D
3. C
4. E
5. B
6. A
7. C
8. D

POPULATIONS
Plant Diversity

GOING GREEN

BIG MOVE

Plants evolved from photosynthetic algae that live in shallow waters. An algal adaptation that helped plants move to land was encasing zygotes in sporopollenin to prevent drying out.

MAKE YOUR OWN LAND PLANT!

WHAT YOU NEED

1. *Growth above and below*

Plants need to absorb light and carbon dioxide from the air and water and nutrients from the soil. **Apical meristems** provide continual growth in the roots and shoots.

2. *Multicellular diploid and haploid stages*

Plants reproduce by **alternation of generations**.

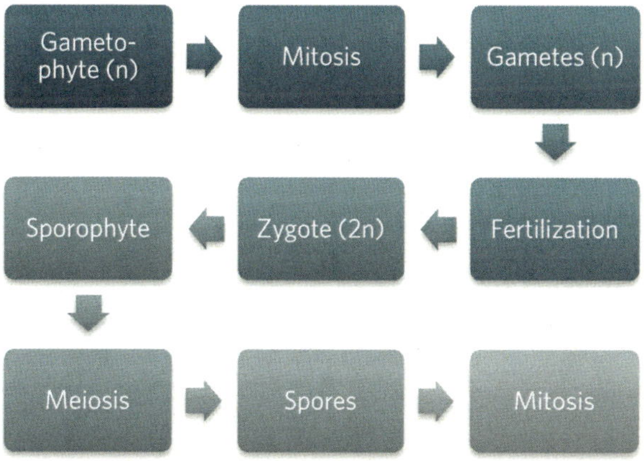

3. *Protected spores*

Plants encase their spores in sporopollenin for protection. Spores develop inside structures called **sporangia**.

4. *Gametes made in multicellular structures*

Plants produce their gametes in **gametangia** (in females) and **antheridia** (in males).

5. *Protected embryos*

Plant embryos develop inside female organs that provide nutrients and protection.

Congratulations!
Read on to find out what kind of plant you made!

CLASS ACTS

FROM FERNS TO FIRS

Plants fall into four categories: vascular (transport nutrients and water through specialized tissues), nonvascular, seed (embryo with protective coat), and seedless.

PLANT PHYLA

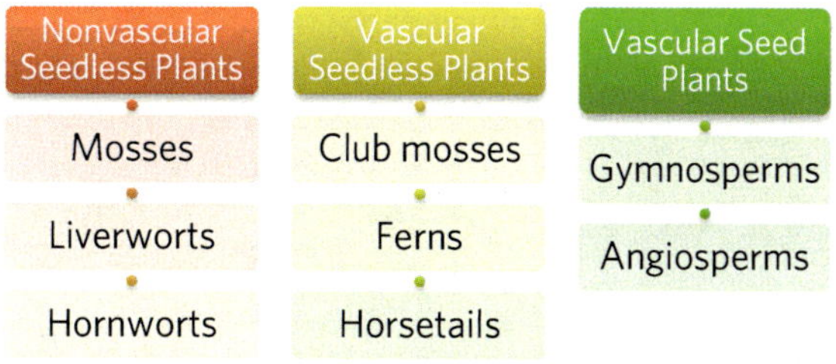

BRYOPHYTES

Mosses, liverworts, and hornworts have older, larger gametophytes and simple sporophytes that absorb nutrients from the parent organism.

LYCOPHYTA/PTEROPHYTA

Ferns, club mosses, horsetails, and other members of this phylum have dominant sporophytes that undergo **homosporous** production (a single spore produces a bisexual gametophyte) and small gametophytes.

SEED PLANTS

Gymnosperms and angiosperms have developed important reproductive adaptations.

GYMNOSPERMS

Ginkgo, cycads, gnetophytes, and conifers have naked seeds that are not enclosed by an ovary but found in modified leaves (cones).

ANGIOSPERMS

Flowering plants encase seeds in mature ovaries (**fruits**), have structures that attract pollinators (**flowers**), and undergo **double fertilization**.

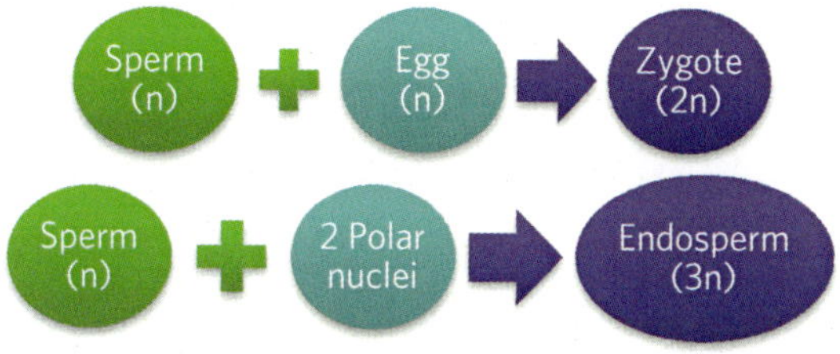

CRAM QUIZ
Plant Diversity

QUESTION 1

What traits do plants and some green algae share that suggest common ancestry?

(A) peroxisomes, cell plate formation, and chitin
(B) sperm without flagella, plasma membrane proteins that produce cellulose, and peroxisomes
(C) flagellated sperm, chloroplasts, and cell plate formation
(D) peroxisomes, flagellated sperm, and plasma membrane proteins that produce cellulose
(E) chloroplasts, sperm without sperm, chitin

QUESTION 2

Traits shared by land plants include

(A) waxy covering on leaves, compounds that absorb UV radiation, and unicellular sporophytes
(B) cuticles, phenolic production, and protected spores
(C) naked seeds, multicellular sporophytes, and cuticles
(D) fruits, alternation of generations, and naked seeds
(E) flowers, spores, and apical meristems

QUESTION 3

What medium do moss sperm use for dispersal?

(A) wind
(B) seta
(C) water
(D) protonemata
(E) antheridia

QUESTION 4

What is the dominant stage in the life cycles of the plants that formed the first forests?

(A) sporophyte
(B) gametophyte
(C) sporangium
(D) archegonium
(E) carboniferous

QUESTION 5

All seed plants share the following traits: pollen, ovules, gametophyte dependent on sporophyte for nutrition, and

(A) homospory
(B) resistant seeds
(C) dominant gamteophyte
(D) heterosporangia
(E) megasporangia and microsporangia

QUESTION 6

Seed production allowed plants to survive in harsh environments. From what tissue is the seed coat derived?

(A) gametophyte
(B) sporangium
(C) integuments
(D) antheridium
(E) heterospory

QUESTION 7

How do gymnosperms and angiosperms differ?

(A) Gymnosperms have independent gametophytes.
(B) Angiosperms have two integuments.
(C) Gymnosperms undergo heterospory.
(D) Angiosperms have dependent sporophytes.
(E) Angiosperms have ovules.

QUESTION 8

In general, what do monocots (one seed leaf) and dicots (two seed leaves) have in common?

(A) root structure
(B) flower structure
(C) vascular tissue arrangement
(D) endosperm structure
(E) leaf venation

ANSWERS

1. D
2. B
3. C
4. A
5. E
6. C
7. B
8. D

POPULATIONS
Animal Diversity

KINGS OF THE JUNGLE

DIFFICULT DEFINITIONS

Animals are not as easily defined as other organisms, but many share traits such as ingesting nutrients, using collagen and certain intracellular junctions for structural support, and having many cells.

SPOTLIGHT ON DEVELOPMENT

Early development in many animals includes cleavage, blastulation, and gastrulation.

- Cleavage
  - Zygote divides mitotically without cell growth
- Blastula Formation
  - Cleavage creates hollow ball of cells
- Gastrulation
  - Cells move and form three tissues (germ layers)

CLASSIFYING THE KINGDOM

Scientists group animals into grades according to body plants and development.

Trait	*Types*
Body Symmetry	▪ Radial ▪ Bilateral
Tissues	▪ Diploblastic ▪ Triploblastic
Body Cavities	▪ Coelomate ▪ Pseudocoalomate ▪ Acoelomate

TYPES OF EARLY DEVELOPMENT

- Protostome
  - Spiral determinant
  - Schizocoelous
  - Mouth formation
- Deuterostome
  - Radial indeterminant
  - Enterocoelous
  - Anus formation

WITH OR WITHOUT A BACKBONE

GO WITH THE FLOW

Animals without backbones (invertebrates) belong to only one phylum but represent 90% of animals. Their members have a variety of special traits.

QUICK REFERENCE: INVERTEBRATES

Animal	*Special Traits*
Sponges	▪ Suspension feeding ▪ Hermaphoditism
Cnidarians	▪ Gastrovascular cavity ▪ Polyp or medusa body plan
Flatworms	▪ Ganglia ▪ Possible parthenogenesis
Molluscs	▪ Visceral mass
Annelids	▪ Segmentation
Nematodes	▪ Cuticle
Arthropods	▪ Exoskeleton ▪ Open circulatory system ▪ Metamorphosis

BACKBONES, BRAINS, AND MORE

Animals with backbones (vertebrates) make up few species but have many strange traits (such as a head with a neural crest (in **crainiates**).

Chordate traits
- Notochord
- Dorsal hollow nerve cord
- Pharyngeal clefts/slits
- Muscular, post-anal tail

QUICK OVERVIEW: VERTEBRATES

- Lampreys
  - Oldest vertebrates
- Sharks
  - Live birth
- Bony fish
  - Swim bladder
- Amphibians
  - Two lives
- Reptiles
  - Extra-embryonic membranes
- Birds
  - Wings/ feathers
- Marsupials
  - Simple placenta
- Placental mammals
  - Complex placenta

CRAM QUIZ
Animal Diversity

QUESTION 1

Bipedalism, brain size, and jaw size in humans evolved at different rates, a phenomenon which is known as

(A) mosaic evolution
(B) punctuated equilibrium
(C) artificial selection
(D) differential evolution
(E) analogy

QUESTION 2

How do craniates and other chordates differ at the genetic level?

(A) Homeotic genes are differentially expressed in crainiates.
(B) Several homeotic gene families are duplicated in crainiates.
(C) They have two groups of *Hox* genes.
(D) They have genes that establish somite formation.
(E) Several gene families are deleted in other chordates.

QUESTION 3

What juvenile features do modern-day chordates retain (a process that is called paedomorphosis)?

(A) hollow nerve cord
(B) anus
(C) somites
(D) notochord
(E) brain

QUESTION 4

In what way do echinoderms and chordates differ?

(A) type of cleavage
(B) sexual reproduction
(C) development of c`oelom
(D) mouth formation
(E) early development

QUESTION 5

When the arthropod's exoskeleton first evolved, its functions included

(A) protection from land animals
(B) flexibility
(C) structural support
(D) prevent drying out
(E) anchoring muscles

QUESTION 6

How do the body plans of triploblastic animals differ?

(A) lining of body cavity
(B) presence of archenteron
(C) number of germ layers
(D) presence of blastocoel
(E) derivation of digestive tract

QUESTION 7

Possible causes of the Cambrian explosion do NOT include

(A) increased atmospheric oxygen
(B) predator-prey interactions
(C) mass extinctions
(D) *Hox* gene evolution
(E) new forms of locomotion

QUESTION 8

When constructing the phylogenic trees of animals, scientists agree that

(A) most animals have radial symmetry
(B) invertebrates and vertebrates are deuterostomes
(C) all animals have true tissues
(D) sponges are eumetazoans
(E) animals are monophyletic

ANSWERS

1. A
2. C
3. D
4. B
5. E
6. A
7. C
8. E

POPULATIONS
Ecology

CLOSE-KNIT COMMUNITY

STUDYING THE WORLD

Ecology examines the complex interactions between organisms and the environment, including nonliving (**abiotic**) and living (**biotic**) factors.

Biotic
- Predation, herbivory, food, parasites, diseases

Abiotic
- Temperature, water, rocks, soil, light, wind

CREATING CLIMATES

Climate represents the weather conditions in a particular area and is influenced by wind, light, temperature, and water.

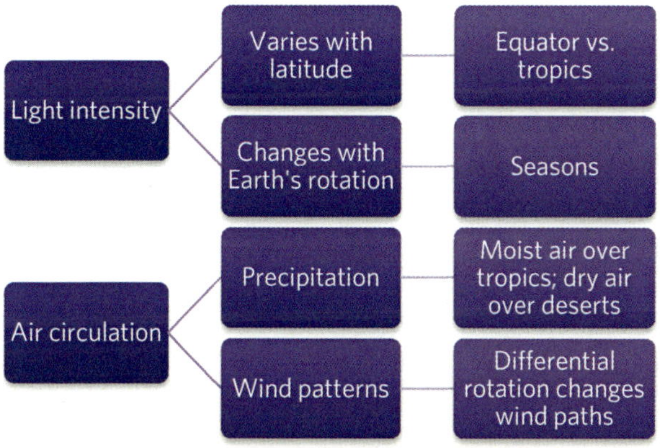

HUMAN IMPACT: GLOBAL CLIMATE CHANGE

What is it?

Fossil fuel and wood combustion sends more carbon dioxide into the atmosphere.

What happens?

Carbon dioxide and water absorb heat and reflect it back to the Earth, raising temperatures. Polar ice caps are melting, and precipitation patterns are changing. Organisms may or may not adapt.

The future?

Unclear until more sophisticated computer modeling programs are available.

BOUNTIFUL BIOMES

ONE IF BY LAND, TWO IF BY SEA

Biomes (either terrestrial or aquatic) are large areas influenced by similar abiotic and biotic factors.

OVERVIEW: TERRESTRIAL/AQUATIC BIOMES

Biome	*Features*
Tropical forest	Rain forests, warm temperatures, near equator, high biodiversity
Desert	Little rain, variable temperatures, low vegetation, small animals
Savanna	Near equator, warm, large herbivores, scattered trees
Chaparral	Coasts, warm summer, seasonal rain, diverse plants
Temperate grassland	Cold winters, hot summers, seasonal rain, grazers, grasses
Coniferous forest	Droughts common, long winters, conifers, diverse birds
Broadleaf forest	Significant rain, humid summers, closed canopy, diverse animals
Tundra	Arctic, short summers, permafrost, dwarf shrubs, grazers
Lake	Standing bodies of water
Wetland	Area covered by water, supports aquatic plants, many plants
Stream/river	Moving bodies of water, diverse fish
Estuary	Area where river and sea meet, many crustaceans and fish
Intertidal zone	Submerged by tide with current, diverse algae
Oceanic pelagic	Open water with currents, phytoplantkton, marine mammals
Marine benthic	Below surface of coastal waters
Coral reef	Tropic marine areas with light, made of calcium carbonate

CRAM QUIZ

Ecology

QUESTION 1

Fertilizers and the cultivation of legumes have increased levels of what nutrient beyond the point where it causes damage to ecosystems (critical load)?

(A) phosphorous
(B) nitrogen
(C) sulfur
(D) calcium
(E) potassium

QUESTION 2

Acid precipitation affects terrestrial ecosystems by

(A) increasing phosphorous beyond critical load
(B) deterioration of bedrock
(C) raising soil pH
(D) causing nutrient leaching into soil
(E) creating acid-tolerant organisms

QUESTION 3

Ozone depletion may have a dramatic effect on what species largely responsible for primary production?

(A) phytoplankton
(B) algae
(C) trees
(D) oomycetes
(E) ferns

QUESTION 4

Lakes and marine environments are divided into zones based on

(A) distance from the bottom
(B) distance from the shore
(C) temperature
(D) water depth
(E) light penetration

QUESTION 5

What aquatic biomes have complex water current patterns?

(A) streams and rivers
(B) intertidal zones
(C) estuaries
(D) ocean pelagic
(E) marine benthic zone

QUESTION 6

What symbiotic organisms provide coral reefs with organic compounds?

(A) brown algae
(B) red algae
(C) plankton
(D) unicellular algae
(E) diatoms

QUESTION 7

What coastal biome is heavily affected by human settlement and agriculture?

(A) chaparral
(B) savanna
(C) temperate grassland
(D) broadleaf forest
(E) coniferous forest

QUESTION 8

Terrestrial biomes feature vertical stratification of what organisms?

(A) animals
(B) insects
(C) plants
(D) microorganisms
(E) decomposers

ANSWERS

1. B
2. D
3. A
4. E
5. C
6. D
7. A
8. C

POPULATIONS
Populations

SIX BILLION AND COUNTING

FRIENDS AND NEIGHBORS

Populations are groups of individuals belonging to the same species, living in the same area, influenced by the same factors, and subject to evolutionary processes.

IN A PERFECT WORLD

A population with unlimited resources grows exponentially according to a specific equation.

MEASURING POPULATION GROWTH

- The difference in birth and death rates is the ***per capita rate of increase*** (r)
- When $r > 0$, a population is growing
- When $r < 0$, a population is shrinking
- Under ideal conditions, a population grows exponentially at a constant rate
- When birth and death rates are equal, the population experiences **zero population growth** (ZPG)

IN THE REAL WORLD

The **carrying capacity** (K) is the maximum number of individuals in a population an environment can support.

- The **logistic population growth** model shows that population growth slows as it reaches carrying capacity

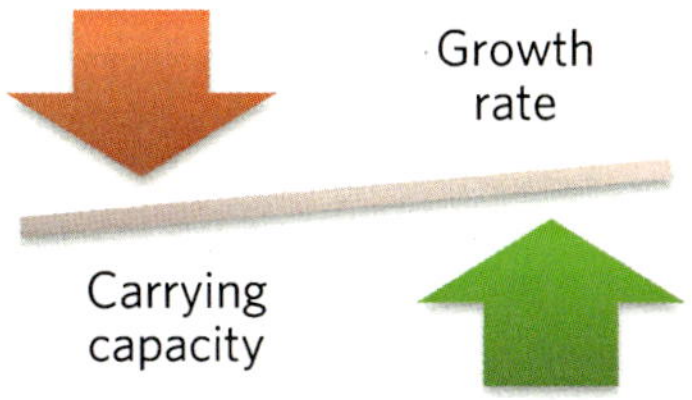

DENSITY MATTERS

Density (the number of individuals per unit area) affects the growth rate of populations, with the exception of **density independent** populations.

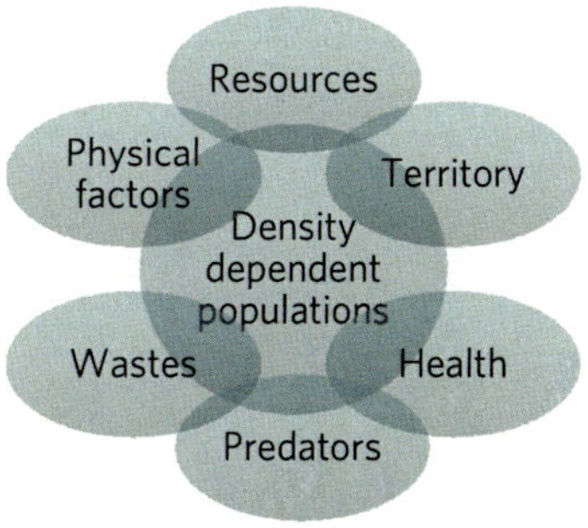

THE KEY TO DIVERSITY

SPECIAL SPECIES

Abundant (**dominant**) or crucial (**keystone**) species greatly impact communities. Dominant species make up the most biomass (mass of total population) in a community. The loss of keystone species can dramatically affect an environment.

HUMAN IMPACT: LOSS OF BIODIVERSITY

What is biodiversity?

Biological diversity occurs at three levels: genetic, species, and ecosystems.

What are the benefits of diversity?

- Lots of products for human use
- Air and water purification
- Detoxifying wastes
- Pollination of plants and crops
- Weather control

What threatens biodiversity?

1. Introduced species
2. Habitat destruction
3. Overuse
4. Disconnected communities

EXAMINING EXTINCTION

Extinction is the irrevocable loss of a species. The **minimum viable population** (MVP) is the smallest survivable population of a species. A small population is pulled closer to extinction by several factors.

THE EXTINCTION VORTEX

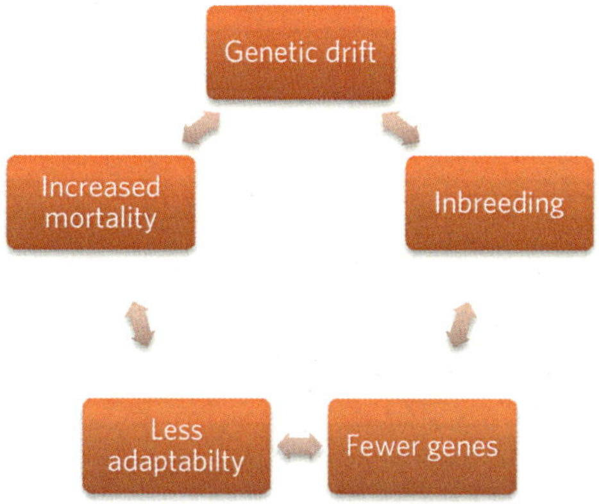

CRAM QUIZ
Populations

QUESTION 1

Fluctuations in size of linked populations (metapopulations) is especially influenced by

(A) birth and death rates
(B) population cycles
(C) immigration and emigration
(D) intrinsic factors
(E) predation

QUESTION 2

In general, low-density populations favor adaptations such as quick reproduction whereas high-density populations favor traits such as

(A) efficient nutrient use
(B) increased growth
(C) predation
(D) more offspring
(E) increased decomposition

QUESTION 3

An organism undergoes semelparity reproduction if

(A) the survival rate of offspring is high
(B) there is fierce competition for resources
(C) the parent is able to care for its young
(D) it lives in an erratic climate
(E) the clutch size is small

QUESTION 4

Inclusive fitness increases an individual's chances of

(A) propagating its genes
(B) its offsprings' survival
(C) being accepted by a community
(D) survival
(E) propagating genes of its close relatives

QUESTION 5

Reciprocal altruism is exhibited by

(A) family members
(B) only humans
(C) symbiots
(D) mates
(E) members of a community

QUESTION 6

The type of mating a species exhibits is closely linked to

(A) caring for young
(B) intrasexual selection
(C) predation
(D) environment
(E) social learning

QUESTION 7

The association of a stimulus to a negative or positive response is

(A) cognition
(B) associative learning
(C) classical conditioning
(D) operant conditioning
(E) spatial learning

QUESTION 8

If a male fish 1) always attacks another male of the same species after 2) seeing markings on its body and 3) chases it away, the fixed action pattern includes

(A) 2 and 3
(B) 3
(C) 1 and 3
(D) 1, 2, and 3
(E) None of these behaviors

ANSWERS

1. C
2. A
3. D
4. E
5. B
6. A
7. C
8. D

POPULATIONS
Communities and Ecosystems

THE WEB OF LIFE

BIOTICALLY BASED

Communities include all the living organisms in a given area close enough to interact. The simplest interactions occur between two species.

COMPETITION

Species compete for a single resource when it is in short supply (**interspecific competition**).

Competition	• In extreme cases, competition eliminates one species (competitive exclusion)
Niche	• All biotic and abiotic resources a species uses
Resource partitioning	• No two species in same place can belong to same niche
Character displacement	• Populations in the same geographic area (sympatric) have different traits and use different resources

PREDATION/HERBIVORY

Several prey species have adapted defenses against predators. Plants may contain toxins, spikes, or thorns to ward off herbivores.

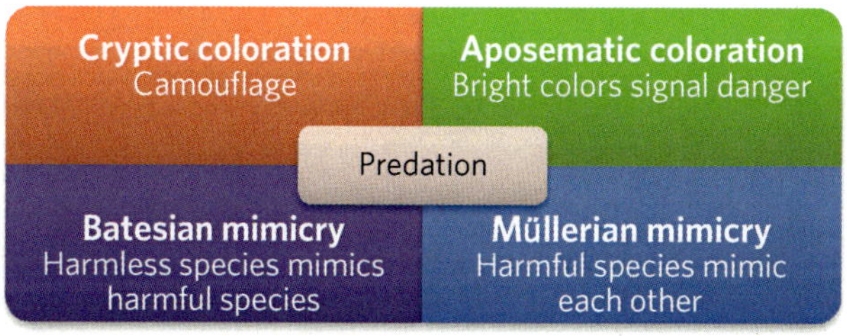

SYMBIOSIS/DISEASE

Interactions between two organisms in direct contact (**symbiosis**) may be beneficial or harmful. **Pathogens** (disease-causing protists, bacteria, viruses, fungi, and prions) are usually microscopic and harmful.

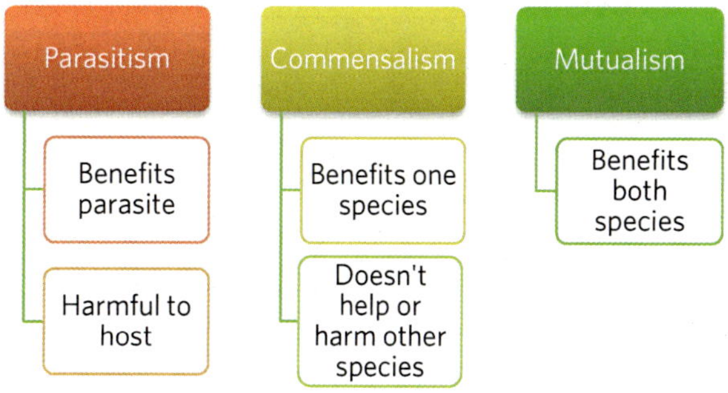

THE WEB OF EVERYTHING

ECOSYSTEMS

Ecosystems include all biotic and abiotic factors affecting a community, with energy cycling through the environment.

TERRIFIC TROPHS

Trophic interactions show the energy sources of each species in an ecosystem. **Biomass** (dry weight of all organisms belonging to the same species) increases as trophic level lowers.

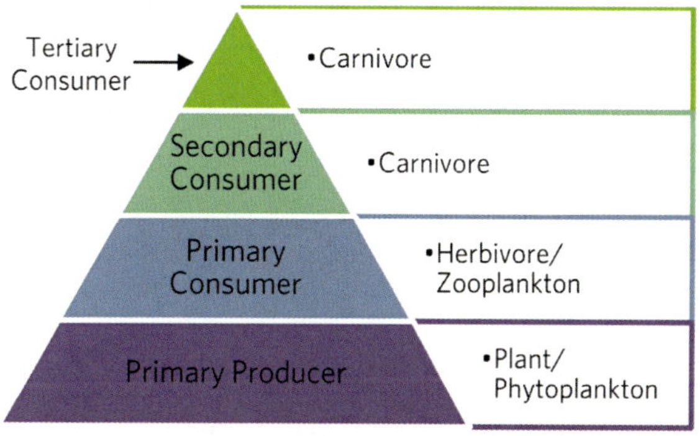

HUMAN IMPACT: BIOLOGICAL MAGNIFICATION

What is it?

Biological magnification is the accumulation of toxins in the food chain.

Consumed biomass increases with higher trophic levels; this process concentrates toxins in top-level consumers. Microbes may not be able to break down certain toxins, or they may convert harmless chemicals to toxins.

ENERGY BUDGETS

Production efficiency is the energy transferred between trophic levels (that isn't used for cellular respiration).

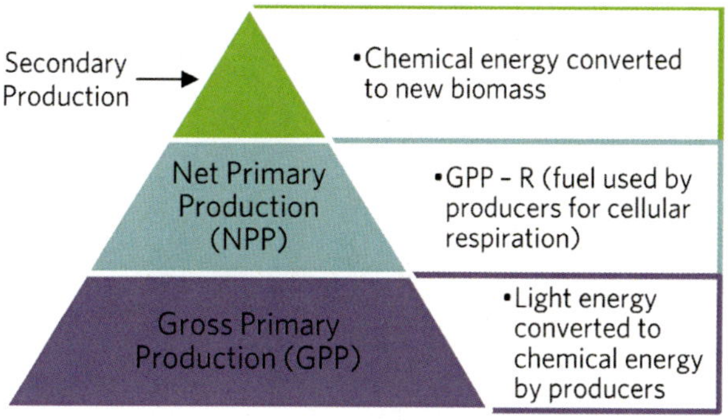

CRAM QUIZ
Communities and Ecosystems

QUESTION 1

Five different communities (One through Five) contain four different species (A through D) in the following ratios. Which community is most diverse? 1) A 25%; B 25%; C 25%; D 25%. 2) A 8%; B 8%; C 74%; D 10%. 3) A 30%; B 20%; C 40%; D 10%. 4) A 50%; B 5%; C 7%; D 38% 5) A 60%; B 7%; C 3%; D 30%

(A) One
(B) Two
(C) Three
(D) Four
(E) Five

QUESTION 2

Food webs consist of food chains of limited length. These limitations may be due to the inefficiency of energy transfer or the

(A) impact of keystone species
(B) impact of decomposers
(C) instability of long food chains
(D) unpredictability of environments
(E) influence of abiotic factors

QUESTION 3

Which model proposes that the predator, rather than the prey, is the primary community organizer?

(A) bottom-up
(A) top-down
(B) energetic
(C) dynamic stability
(D) intermediate disturbance

QUESTION 4

Why are tropical habitats more diverse than polar or temperate habitats?

(A) Polar and temperate ecosystems are older.
(B) The growing season in polar and temperate ecosystems is longer.
(C) Polar and temperate climates have had fewer disturbances.
(D) Tropic areas have greater actual evapotranspiration.
(E) Tropic areas have lower potential evapotranspiration.

QUESTION 5

Aquatic ecosystems may have inverted biomass pyramids because

(A) phytoplankton have a small standing crop biomass
(B) phytoplankton reproduce quickly
(C) zooplankton's trophic efficiency is high
(D) zooplantkton's turnover time is short
(E) zooplankton have a higher production

QUESTION 6

Aquatic ecosystems have slow rates of decomposition due to

(A) increased buildup of detritus
(B) increased nutrient exchange between surface and bottom layers of water
(C) high rates of evapotranspiration
(D) low concentrations of nutrients in the sediment
(E) algae and aquatic plants taking nutrients from the water

QUESTION 7

Movement corridors are often used to

(A) remove edge communities
(B) change habitat composition
(A) connect fragmented habitats
(B) encourage the growth of edge-adapted populations
(C) slow down the spread of disease

QUESTION 8

Nature reserves are established to

(A) avoid ecological disturbances
(B) create an ecosystem according to the nonequilibrium model
(C) create a balanced environment that self-regulates
(D) preserve disturbances and population dynamics
(E) create large areas of preserved land

ANSWERS

1. A
2. C
3. B
4. D
5. A
6. E
7. C
8. D

PLANTS
Plant Structure

GREEN MACHINES

ROOTS TO SHOOTS

Plants' three organs (leaves, stems, and roots) are well built for taking nutrients and water from the ground, carbon dioxide from the air, and light energy from the sun.

ROOTS

Root systems anchor the plant in the ground and absorb nutrients and water from the soil through specialized cells that increase surface area (**roots hairs**).

- **Taproot** systems have a main vertical root (which may store food) and many smaller branches
- **Fibrous** root systems have thinner roots that branch out shallowly beneath the soil

STEMS

Stems transport water and nutrients to the leaves and sugar to the roots. Leaves attach to stems at **nodes**, with **internodes** in between. At each node, an **axillary bud** may form a branch but is inhibited by the shoot tip (**terminal bud**).

LEAVES

Leaves are responsible for gas exchange and sugar production. A wide **blade** attaches to the stem via a stalk (**petiole**).

PLANT TISSUES

Each plant organ has three types of tissue. Dermal tissue protects the plant and prevents water loss. Vascular tissue transports nutrients and water, and ground tissue includes various support cells.

Dermal	Epidermis	Single layer cells; Nonwoody plants
	Periderm	Replaces epidermis; Woody plants
	Cuticle	Waxy coating Stems and leaves
Vascular/ Stele	Xylem	Water and nutrient transport
	Phloem	Sugar transport
Ground	Pith	Interior to vascular
	Cortex	Exterior to vascular

BELOW THE SURFACE

TASK MASTERS

Plant cells are specialized to perform certain tasks but fall into different categories based on their structure and function.

THE FANTASTIC FIVE

Plants cells are grouped into five categories, which may be further divided.

Parenchyma	•Least specialized •Many functions •Totipotent
Sclerenchyma	•**Sclerids** •Thick, hard walls •**Fibers** •Slender, strong
Collenchyma	•Support a young plant •Uneven thickness •Structurally flexible
Water-conducting	•**Tracheids** •Prevent collapse; slender, tapered •**Vessel elements** •Form vessels; wider, thinner
Sugar-conducting	•**Sieve-tube members** •no organelles; transport nutrients •**Companion cells** •Support sieve-tube

ANATOMY OF A LEAF

Leaves maximize gas exchange and evaporation by layering different types of tissue.

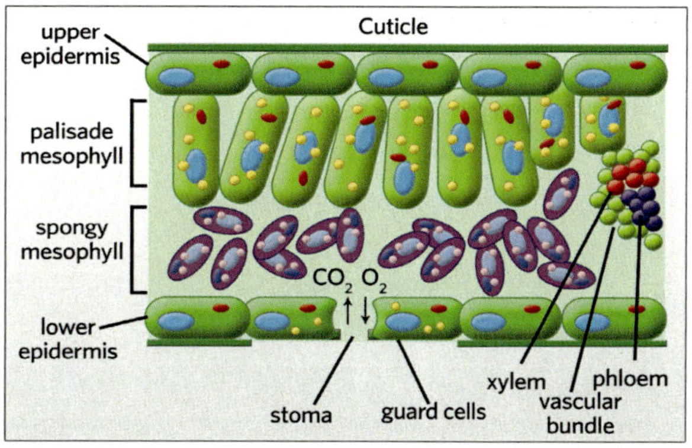

CRAM QUIZ
Plant Structure

QUESTION 1

In the roots of gymnosperms, the stele is arranged
(A) with a core of xylem
(B) similarly to most monocots
(C) with parenchyma cells in the center
(D) with a ring of xylem
(E) with interspersed xylem and phloem

QUESTION 2

Roots hair grow out from the epidermis in the
(A) root cap
(B) zone of division
(C) zone of elongation
(D) zone of maturation
(E) entire root

QUESTION 3

Apical dominance allows a plant to
(A) conserve resources
(B) withstand adverse conditions
(C) encourage cell division
(D) maximize light exposure
(E) outcompete other plants

QUESTION 4

What structure(s) may be modified for food storage?
(A) leaves
(B) stems and roots
(C) stems
(D) leaves and roots
(E) roots

QUESTION 5

Plant cells that do NOT contain lignin include
(A) parenchyma
(B) collenchyma
(C) parenchyma and sclernchyma
(D) sclernchyma and collenchyma
(E) parenchyma and collenchyma

QUESTION 6

In tracheids, water flows from cell to cell through
(A) plasmodesmata
(B) gap junctions
(C) primary cell walls
(D) vessels
(E) companion cells

QUESTION 7

In sieve-tube members, organic nutrients travel from cell to cell through
(A) sieve plates
(B) gap junctions
(C) primary cell walls
(D) companion cells
(E) plasmodesmata

QUESTION 8

In leaves, the palisade mesophyll is made up of what type of tissue?
(A) pith
(B) epidermis
(C) collenchyma
(D) cortex
(E) periderm

ANSWERS

1. A
2. C
3. D
4. B
5. E
6. C
7. A
8. D

PLANTS
Plant Growth

MAGICAL MERISTEMS

IT KEEPS GROWING, AND GROWING . . .

Plants grow throughout their lifetimes due to embryonic tissues called **meristems**. These cells may become specialized (**derivatives**) or continue to produce more embryonic cells (**initials**).

ROOT GROWTH

An apical meristem covered by a protective root cap continually adds new growth the roots.

Zone of maturation
- Differentiated cells
- Root hairs

Zone of elongation
- Pushes root into soil

Zone of division
- Meristem and derivatives

Root cap
- Secretes lubricant
- Protects meristem

DIVIDE AND CONQUER

Plant cells divide along planes of symmetry or they may expand in size, depending on the cells' functions.

Division	Expansion
Preprophase band forms in late interphase	Enzymes break down cell wall
Plane of symmetry determined	Water flows into cell
Cell divides in single file, into a cube, or assymetrically	Cell expands 10 times its size

TRICKY TRANSITIONS

Plant regions, as opposed to entire organisms, pass from one life phase to another, thanks to meristem growth.

Juvenile → Adult vegetative → Adult reproductive

GROWING UP (AND OUT)

STRAIGHT SHOOTER

The terminal bud, a dome of cells at the tip of a shoot, is the site of primary growth. It is flanked by leaf primordial (site of leaf growth) and axillary bud meristems.

ADDING GIRTH

The thick stems and branches of woody plants are secondary growth tissues made by the vascular and cork cambium.

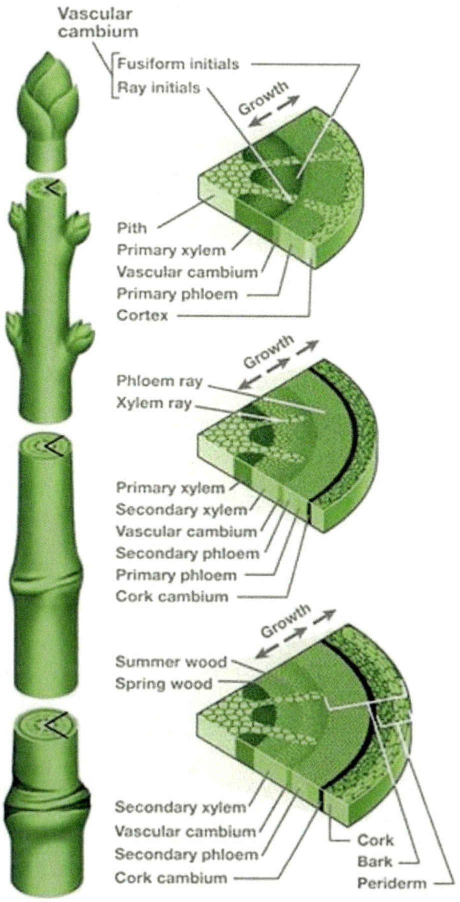

A CLOSER LOOK: SECONDARY GROWTH

1. Fusiform initials form secondary xylem and phloem; ray initials form xylem and phloem rays.
2. Tissues around vascular cambium rupture and are replaced by cork cells.
3. Cells outside cork cambrium rupture and fall off.
4. Cork cambium continues to replace cortex and form periderm.
5. Bark forms from tissues outside of vascular cambium.

CRAM QUIZ
Plant Growth

QUESTION 1

Lateral roots grow out from what to remain continuous with the plant's vascular system?

(A) endodermis
(B) epidermis
(C) pericycle
(D) apical meristem
(E) vascular tissue

QUESTION 2

Place the tree trunk tissues in the correct order from exterior to interior: 1) sapwood 2) periderm 3) heartwood 4) vascular cambium 5) secondary phloem.

(A) 1, 3, 5, 4, 2
(B) 3, 2, 4, 5, 1
(C) 2, 4, 5, 3, 1
(D) 2, 5, 4, 1, 3
(A) 3, 4, 2, 1, 5

QUESTION 3

In monocot stems,

(A) pith and cortex are separate
(B) vascular bundles are interspersed
(C) there are sclerenchyma cells
(D) vascular bundles have no sheath
(E) xylem and phloem are separate

QUESTION 4

What is the organization of leaf tissue in order from the top to the underside of the leaf?

(A) cuticle, epidermis, palisade mesophyll, spongy mesophyll, epidermis, cuticle
(B) stomata, epidermis, palisade mesophyll, spongy mesophyll, stomata, cuticle
(C) cuticle, epidermis, spongy mesophyll, palisade mesophyll, epidermis, cuticle
(D) cuticle, stomata, palisade mesophyll, spongy mesophyll, stomata, cuticle
(E) cuticle, stomata, spongy mesophyll, palisade mesophyll, stomata, cuticle

QUESTION 5

What cells are oriented parallel to the stem's axis and form elongated cells such as companion cells?

(A) vascular cambium
(B) ray initials
(C) collenchyma cells
(D) phloem
(E) fusiform initials

QUESTION 6

What function do lenticels perform?

(A) prevent water loss
(B) cell division
(C) gas exchange
(D) stem protection
(E) structural support

QUESTION 7

Organ identity genes are plant homeotic genes that regulate the

(A) positioning of flower structures
(B) growth of plant organs
(C) preprophase plate formation
(D) indeterminate growth of meristems
(E) differentiation of stems

QUESTION 8

Early spring stem growth

(A) adds support to the plant
(B) increases water transport
(C) is known as sapwood
(D) adds less secondary xylem and more phloem
(E) has narrower cells

ANSWERS

1. C
2. D
3. B
4. A
5. E
6. C
7. A
8. B

PLANTS
Vascular Transport

AGAINST THE GRAIN

VARIABLE VESSELS

Vascular plants transport nutrients and water through a system of vessels that crisscross the entire plant and transfer materials among cells, tissues, and organs.

SIDE TO SIDE: LATERAL TRANSPORT

Roots absorb water and nutrients from the soil and transfer them to the rest of the plant through a process called lateral transport.

Choose Your Route

Materials enter roots via three different routes, some of which are more selective.

1. **Apoplast**—the extracellular spaces between cells
2. **Symplast**—plasmodesmata that connect cells
3. **Transmembrane**—across plasma membranes

The **Casparian strip** is a selective barrier that forces transport through the symplast.

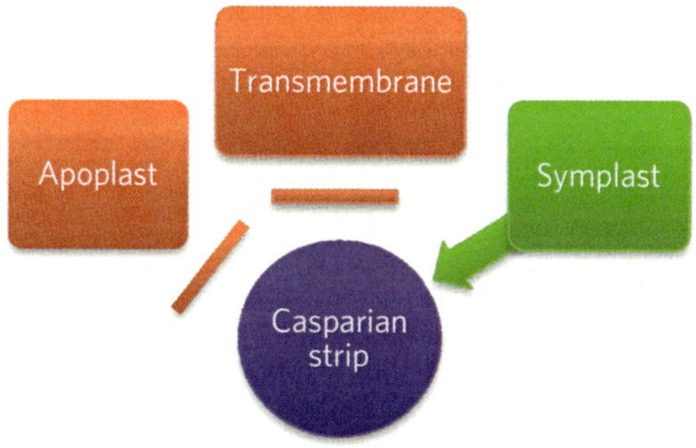

BOTTOMS UP: BULK FLOW

Several factors cause water and dissolved nutrients to move against gravity up the stem. Overall, pressure drives vascular transport in a process called bulk flow.

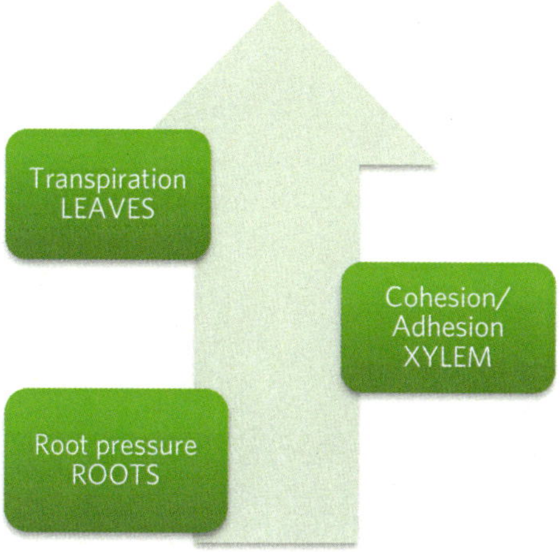

GOING UP, COMING DOWN

THE SKINNY ON TRANSPORT

Differences in **water potential** (measure of solute concentration and pressure) pull xylem sap up and phloem sap down the plant.

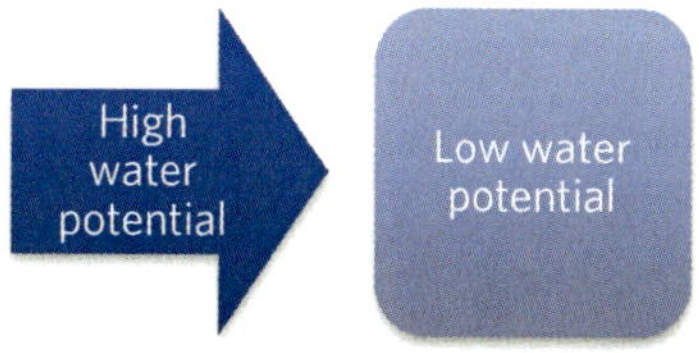

MOVING THE SAP

The entire plant is involved in the movement of xylem sap, the water and dissolved nutrients transported by xylem tissue.

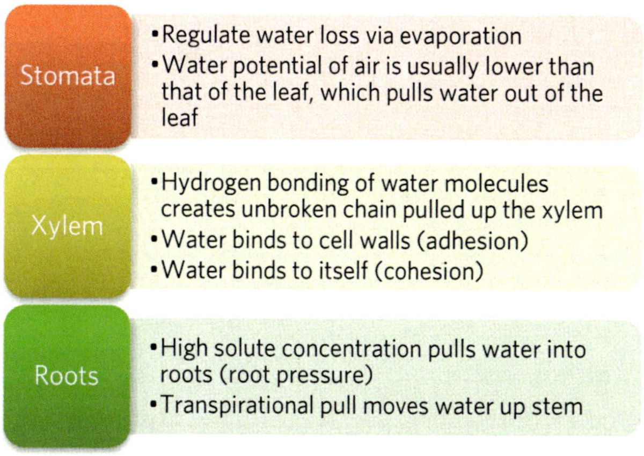

MOVING THE SUGAR

Phloem sap contains sugars, amino acids, hormones, minerals, and water. It moves from **source** (where sugar is produced) to **sink** (where sugar is used or stored).

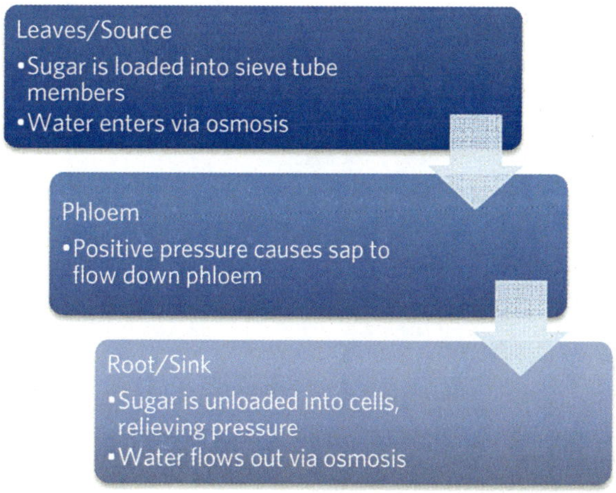

CRAM QUIZ
Vascular Transport

QUESTION 1

What allows plant cells to drive the transport of many different solutes?

(A) stored energy from proton gradients
(B) ATP pumps
(C) membrane proteins coupled with ATP syntase
(D) diffusion
(E) water potential

QUESTION 2

Anions are usually transported into the cell via what mechanism?

(A) differences in membrane potential
(B) anion uptake
(C) cotransport
(D) diffusion
(E) active transport

QUESTION 3

How do pressure potential (Ψ_p) and solute potential (Ψ_s) relate to water potential (Ψ)?

(A) $\Psi_p - \Psi_s = \Psi$
(B) $\Psi_s - \Psi_p = \Psi$
(C) $\Psi + \Psi_s = \Psi_p$
(D) $\Psi + \Psi_p = \Psi_s$
(E) $\Psi_p + \Psi_s = \Psi$

QUESTION 4

How does the pressure potential of xylem cells and other living plants cells differ?

(A) Xylem cells experience turgor pressure.
(B) Living cells are under positive pressure.
(C) The pressure of xylem cells is always a negative number.
(D) The pressure of other cells is negative compared to the atmosphere.
(E) Other cells experience tension.

QUESTION 5

Aquaporins affect water transport across membranes by

(A) changing the direction of movement
(B) inhibiting movement
(C) affecting water potential
(D) enhancing movement
(E) changing osmotic potential

QUESTION 6

Once xylem sap enters the vascular cylinder, it moves from

(A) casparian strip to apoplasts to vessels
(B) casparian strip to symplasts to vessels
(C) parenchyma cells to symplasts to vessels
(D) apoplasts to parenchyma cells to vessels
(E) parenchyma cells to apoplasts to vessels

QUESTION 7

The opening and closing of guard cells is regulated by

(A) differences in pressure potential
(B) changes in transpiration rates
(C) the accumulation and loss of potassium ions
(D) stomata
(E) neighboring mesophyll cells

QUESTION 8

The transport of sugar into the sieve-tube members and companion cells is driven by

(A) chemiosmosis
(B) diffusion
(C) proton pumps
(D) passive transport
(E) cotransporters

ANSWERS

1. A
2. C
3. E
4. B
5. D
6. E
7. C
8. A

PLANTS
Angiosperm Reproduction

GOOD BREEDING

SWEET-SMELLING SEX ORGANS

Flowers are sophisticated sex organs molded by adaptation to increase reproductive success.

COLORFUL ADAPTATIONS

A flower is a specialized structure designed to increase genetic diversity.

Stamen (male organ)—filament, anther, pollen

Carpel (female organ)—ovary, ovules, stigma, style

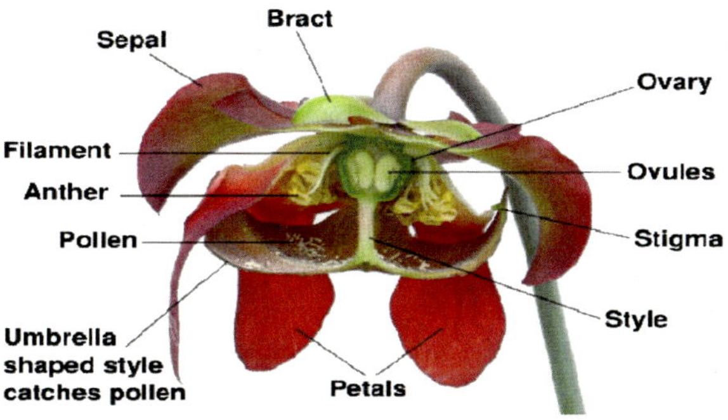

GROWING GAMETES

Male gametophytes (**pollen**) develop inside the anthers (microsporangia). Female gametophytes (**embryo sacs**) develop inside the ovule, which is surrounded by the ovary.

Pollen	Embryo sac
Microsporocyte (2n) in pollen sac	Megasporocyte (2n) inside megasporagium
Meiosis	Meiosis
Four microspores (n)	Four megaspores (n), one survives
Mitosis	Three mitotic divisions
Two generative and one tube cell (pollen)	Three antipodal cells, two polar nulcei, one egg, two synergids surrounded by integuments (embryo sac)

FROM POLLEN TO PLANT

PROTECTIVE PARENTS

Angiosperms reproduce twice to create new plants and nurture them in a process called **double fertilization**.

DOUBLE OR NOTHING

Double fertilization creates a diploid zygote and the nutritive tissue (**endosperm**) surrounding it.

Pollen tube
- Pollen lands on stigma
- Tube cell produces pollen tube
- Tube grows down stigma
- Generative cell makes two sperm

Sperm discharged
- Two sperm enter ovule through micropyle

Double fertilization
- Sperm + egg = zygote
- Sperm + polar nuclei = endosperm

PLANT LIFE: AN OVERVIEW

1. The zygote develops into an embryo.
2. The embryo has simple roots (radicle), stems (hypocotyl), and a shoot tip (epicotyl) encased in a layer of integuments that harden to form a seed coat.
3. The seed dehydrates and enters dormancy.
4. The seed takes in water (imbibition) and germinates.
5. The hypocotyl breaks ground and grows as the epicotyl's leaves mature.
6. The seedling matures until it reaches the reproductive phase.

Zygote → Mitosis → Embryo → Seed maturity → Dormancy → Germination → Growth → Maturity → Reproduction

WITHOUT SEX

Plant fragments reproduce **vegatatively** (clonally) thanks to meristematic tissues. Ovules can produce seeds asexually in a process known as **apomixis**.

CRAM QUIZ
Angiosperm Reproduction

QUESTION 1

What is the first embryonic structure to emerge from the seed?

(A) epicotyl
(B) hypocotyl
(C) radicle
(D) cotyledon
(E) seed coat

QUESTION 2

Which of the following characteristics applies to the endosperm?

(A) unicellular with many nuclei
(B) without cell walls
(C) transfers nutrients to radicle
(D) exchanges nutrients with cotyledons
(E) is always liquid

QUESTION 3

From which of the following structures do corn seeds tend to absorb nutrients?

(A) the endosperm
(B) a single cotyledon
(C) a pair of cotyledons
(D) the coleoptile
(E) the scutellum

QUESTION 4

What is the difference between sporophyte and gametophyte self-incompatibility?

(A) Gametophyte self-incompatibility involves signal transduction pathways.
(A) Gametophyte self-incompatibility causes the hydrolysis of sperm RNA.
(B) Sporophyte self-incompatibility blocks fertilization.
(C) Sporophyte self-incompatibility hydrolyzes DNA of sperm.
(D) Gametophyte self-incompatibility stops pollen from germinating.

QUESTION 5

What is responsible for the development of an ovary into a fruit?

(A) external stimuli
(B) sugar production
(C) seed maturation
(D) water
(E) hormones

QUESTION 6

Which of the following mechanisms in angiosperms most commonly prevents self-fertilization and increases genetic diversity?

(A) blocking pollen development
(B) vegetative reproduction
(C) separation of male and female gametophytes by dioecious plants
(D) changing maturation times of gametophytes
(E) blocking apoximis

QUESTION 7

What are the functions of the three antipodal cells and the two synergids of the female gametophyte, respectively?

(A) unknown; forming endosperm
(B) forming seed coat with the integuments; forming endosperm
(C) unknown; guiding pollen tube
(D) guiding pollen tube; forming endosperm
(E) forming seed coat; guiding pollen tube

QUESTION 8

What happens in a seed immediately after imbibition?

(A) nutrients diffuse into the embryo
(B) enzymes break down food reserves
(C) water forces the radicle to emerge
(D) the seed coat dissolves
(E) cotyledons break down

ANSWERS

1. C
2. D
3. A
4. B
5. E
6. A
7. C
8. B

PLANTS
Plant Hormones

FROM DEFENSE TO DIFFERENTIATION

COMPLICATED CUES

A variety of chemical signals called hormones coordinate cell activities by binding to receptors on target cells and changing cell processes.

QUICK REFERENCE

Several different hormones regulate plant functions.

Hormone	Major Functions
Auxin	Elongation, root growth, differentiation
Ethylene	Ripening, leaf loss, mechanical stress
Brassinosteroids	Differentiation, slow leaf loss
Abscisic acid	Inhibit growth, close stomata, seed dormancy
Cytokinins	Division, differentiation, apical dominance, retard aging
Gibberellins	Stimulate germination, stem elongation, fruit growth

PERFECT PARTNERSHIPS

Hormones often work in concert to produce a desired effect.

Hormones	Interactions
Auxin + cytokinins	▪ Auxin = elongation ▪ Cytokinins = division ▪ Auxin + cytokinins = differentiation ▪ High auxin = roots ▪ High cytokinins = shoots
Auxin + ethylene	▪ Auxin + ethylene = regulation of leaf loss ▪ Low auxin = lose leaves
ABA + gibberellins	▪ High gibberellin = germination ▪ High ABA = dormancy

UNDER THE MICROSCOPE

AUXIN AND CELL ELONGATION

Where is auxin?

In the embryo, immature leaves, and apical buds.

What does it do?

Auxin may cause cell elongation by turning on proton pumps in the plasma membrane, which is called the acid growth hypothesis.

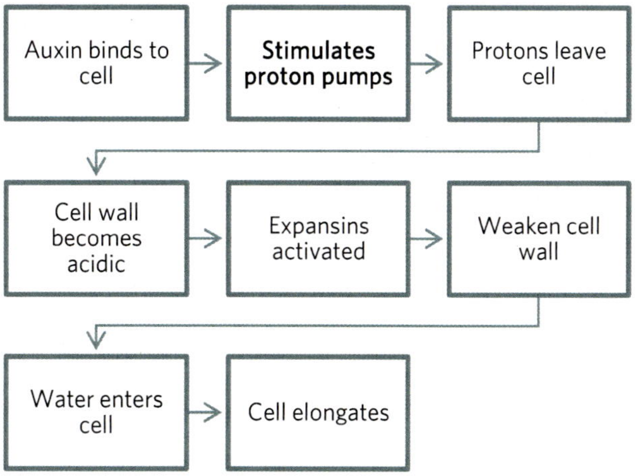

CYTOKININS AND CELL DIVISION

Where are cytokinins?

In growing tissues—roots, embryos, and fruits.

What do cytokinins do?

- When only cytokinin is present, the cells do not divide
- When cytokinin and auxin are both present, the cells divide but are undifferentiated
- When the ratio of auxin to cytokinin changes, the cells differentiate

ETHYLENE AND RIPENING

Where is ethylene?

In fruits, nodes, old leaves, and flowers.

What does it do?

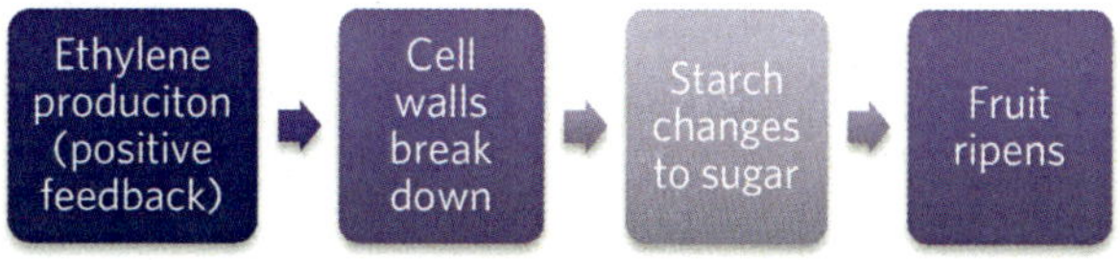

CRAM QUIZ
Plant Hormones

QUESTION 1

The triple response of ethylene to mechanical stress begins with

(A) decrease in cell elongation
(B) increase in cell elongation on one side of the stem
(C) stem thickening
(D) stem curvature
(E) decrease in cell elongation on one side of the stem

QUESTION 2

Ethylene production is usually associated with cell

(A) growth
(B) elongation
(C) differentiation
(D) inhibition
(E) death

QUESTION 3

Wilting triggers the production of what hormone that triggers the opening of potassium channels of guard cells?

(A) ethylene
(B) auxin
(C) ABA
(D) gibberellin
(E) cytokinin

QUESTION 4

Gibberellins may cause stem elongation by

(A) acidifying the cell wall
(B) breaking down the cytoskeleton
(C) opening aquaporins
(D) affecting expansins
(E) cleaving polysaccharides

QUESTION 5

In the direct inhibition hypothesis, auxin and cytokinin interact in what way to regulate apical dominance?

(A) Auxin transported from roots stimulates apical buds.
(B) Auxin transported from shoots inhibits apical buds.
(C) Cytokinins transported from roots inhibit terminal bud.
(D) Cytokinins transported from shoots inhibit terminal bud.
(E) Auxin transported from roots inhibits apical buds.

QUESTION 6

Seedless fruits are grown with the application of synthetic

(A) ABA
(B) ethylene
(C) cytokinin
(D) gibberellin
(E) auxin

QUESTION 7

Auxin causes cell elongation by

(A) causing expansins to penetrate cell walls
(B) activating transcription factors
(C) affecting gene expression
(D) changing post-translational modifications of proteins
(E) stimulating transcription

QUESTION 8

A coleptile bends toward light under what conditions?

(A) when it is covered by a transparent cap
(B) when the tip is removed
(C) when mica is placed underneath the tip
(D) when agar is placed underneath the tip
(E) when the tip is replaced by agar soaked in auxin

ANSWERS

1. A
2. E
3. C
4. D
5. B
6. E
7. C
8. A

PLANTS
External Stimuli and Defense

STIMULATING THE SENSES

BEING RESPONSIVE

Plants often respond quickly and efficiently to their environments—an important adaptation for immobile organisms.

LIGHT OF MY LIFE

Response to light is crucial to plant survival. **Photoreceptors** detect light intensity, direction, and color and produce different effects.

- Blue light receptors respond to blue light
- Phytochromes respond to red light

Blue Light Photoreceptors		
Cryptocrome	Phototropin	Zeaxanthin
•Slows hypocotyl elongation	•Bends plant toward light	•Opens stomata

ISOMETRIC INTERACTIONS

Phytochrome exists in two different forms—P_r and $P_{fr.}$ Far red light changes P_{fr} to $P_{r,}$, and red light reverses the transformation.

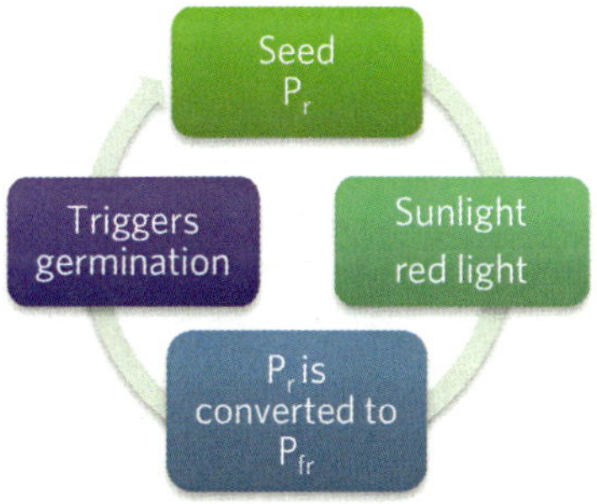

SPOTLIGHT ON SAR

Systemic acquired resistance (SAR) is one way a plant defends itself against pathogens.

1. The pathogen binds to plant cells.
2. The plant has a resistance (*R*) gene for that pathogen.
3. The pathogen is avirulent (causes mild harm).
4. The signal transduction pathway is triggered.
5. The cell destroys itself via hypersensitive response.
6. The infected area is sealed, and lesions form.
7. Infected cells release a chemical (probably salicylic acid).
8. The signal travels throughout the plant.
9. Protection compounds are produced via SAR response.
10. Healthy cells receive added protection.

MAGNIFICENT MANUVERS

SIGNIFICANT STIMULI

Plants respond to environmental stimuli in creative ways to make up for their immobility.

BENDING BEAUTIFULLY

Gravity, touch, and harsh conditions affect plant responses in many ways, which are often called **tropisms**.

Gravity	•Statoliths move •Shoots grow against gravity •Roots grow with gravity
Touch	•Calcium ions flood cytosol •Change direction of growth; stunt growth
Flooding	•Ethylene causes apoptosis •Air tubes form in roots
Heat	•Heat shock proteins made •Prevent denaturation
Freezing	•Plasma membrane changes lipid composition

SHOW NO MERCY

SECRET WEAPONS

Plant defenses may not be visible, but they are formidable, deterring predators and pathogens.

CRAM QUIZ

External Stimuli and Defense

QUESTION 1

The function(s) of the two domains of each phytochrome protein is(are)

(A) to respond to light
(B) to change phytochrome conformation and respond to light
(C) to change phytochrome conformation
(D) to activate proteins and respond to light
(E) to bond covalently to other phytochrome proteins

QUESTION 2

How do the two isomers of phytochrome affect plant growth?

(A) P_r encourages branching.
(B) P_r inhibits elongation.
(C) P_r stimulates vertical growth.
(D) P_{fr} increases in shade.
(E) P_{rf} causes shade avoidance.

QUESTION 3

Phytochrome regulates circadian rhythms by

(A) rapidly increasing the concentration of one isomer at sunrise
(B) rapidly decreasing the concentration of one isomer at sunset
(C) slowly degrading one isomer at night
(D) quickly degrading one isomer at sunset
(E) keeping the isomers at equilibrium

QUESTION 4

A long day plant requiring vernalization won't flower until it is exposed to

(A) several successive days of uninterrupted darkness
(B) cold and uninterrupted darkness
(C) an external stimulus and several long days of sunlight
(D) an external stimulus and darkness exceeding the critical night length
(E) darkness exceeding the critical night length

QUESTION 5

Gravitropism may be caused by

(A) release of auxin by statoliths
(B) stimulation of cell elongation
(C) lateral transport of calcium into cells
(D) decrease in auxin production
(E) accumulation of auxin and calcium

QUESTION 6

How do extremely touch-sensitive plants respond to mechanical stimuli?

(A) very fast action potentials
(B) activation of proton pumps
(C) redistribution of calcium ions
(D) via motor organs
(E) by gaining turgor pressure

QUESTION 7

A pathogen does not harm a plant if

(A) the plant has an *R* allele and the pathogen has no *Avr* allele
(B) the plant has no *R* allele and the pathogen has an *Avr* allele
(C) the plant has an *R* allele and the pathogen has an *Avr* allele
(D) the pathogen has no *Avr* allele and the plant has no *R* allele
(E) the plant has a similar *R* allele and the pathogen has an *Avr* allele

QUESTION 8

Elicitors mount generalized defenses against pathogens by

(A) depositing lignin in cell walls
(B) producing antifungal compounds
(C) stimulating SAR response
(D) killing infected cells
(E) activating genes that produce phytoalexins

ANSWERS

1. D
2. C
3. A
4. B
5. E
6. D
7. C
8. A

ANIMALS
Animal Structures

AMAZING ANATOMY

HECTIC HIERARCHY

Animals are incredibly diverse. Their cells are often specialized and organized into increasingly complex systems that are molded to a particular function.

THE FANTASTIC FOUR

Animal tissues fall into four groups, based on structure and function.

Quick Reference

Tissue	*Function*
Epithelial	Protects outside and inside of body
Connective	Connects other tissues
Muscle	Cells that contract
Nervous	Responds to stimuli

A CLOSER LOOK AT TISSUES

Epithelial

Tightly packed cells line the body cavity and form the skin.

Shapes: Cuboidal, Columnar, Squamous

Layers: Single, Stratified, Pseudo-stratified

Connective

Cells scattered in a network of embedded fibers support other tissues.

Cells
- Fibroblasts
  - Secrete proteins
- Macrophages
  - Defense/cleanup

Fibers
- Collagenous
  - Prevents tearing
- Elastic
  - Elasticity
- Reticular
  - Tight network

Tissues
- Loose
- Adipose
- Fibrous
- Cartilage
- Bone
- Blood

SUPERB SPECIALIZATION

A CLOSER LOOK AT TISSUES (CONTINUED)

Muscle

Muscle is the most abundant tissue.

- Skeletal
  - Voluntary movement
- Smooth
  - Involuntary movement
- Cardiac
  - Heart contraction

Actin and myosin make up long, parallel fibers that contract when stimulated.

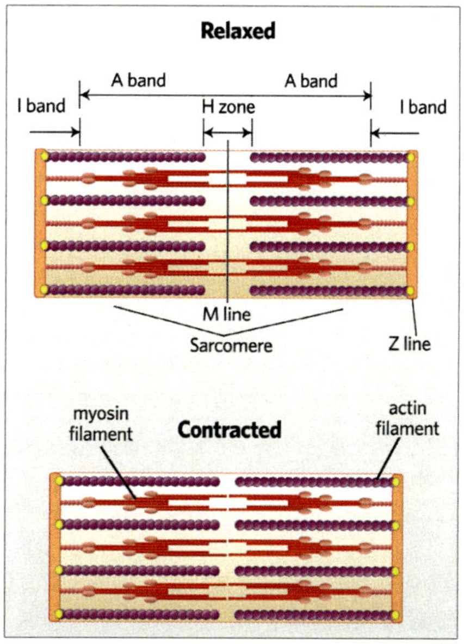

Nervous

Nervous cells (neurons) are built to receive and transmit electrical impulses that respond to stimuli.

Axon: Receives signal → Cell body: Transfers signal → Dendrite: Transmits signal

The brain is the nervous system's control center with different specialized regions.

Brain structure	*Function*
Brainstem	Sleep, arousal, homeostasis, coordination, relaying information
Cerebellum	Coordination, cognitive functions, motor memory
Diencephalon	Circadian rhythms, sensory/motor processing, mating, stress response, control basic survival
Cerebrum	Language, motor activities initiated, sensory processing

CRAM QUIZ
Animal Structures

QUESTION 1

The skeleton combines the flexibility of collagen with the strength of

(A) layers of fibrous tissue
(B) osteoblasts
(C) calcium, magnesium, and phosphate ions
(D) osteons
(E) various minerals

QUESTION 2

The cartilaginous skeletons of embryonic organisms contain collagenous fibers and

(A) chondrocytes
(B) a protein-carbohydrate network
(C) hydroxyapatite
(D) chondroitin
(E) osteons

QUESTION 3

Stratified squamous epithelial tissue is organized in such a way to ensure that

(A) old cells are worn away
(B) nutrients can diffuse
(C) foreign particles are trapped and removed
(D) nutrients are absorbed
(E) chemicals are secreted

QUESTION 4

What is the difference between skeletal and cardiac muscle?

(A) Cardiac muscle lacks striations.
(B) Skeletal muscle contracts more slowly.
(C) Cardiac muscle fibers branch.
(D) Skeletal muscle contains sarcomeres.
(E) Skeletal muscle is multinucleated.

QUESTION 5

Heat is lost during what cellular processes?

(A) consumption and cellular respiration
(B) organic molecule synthesis and reproduction
(C) excretion and organic molecule synthesis
(D) digestion and consumption
(E) digestion and cellular respiration

QUESTION 6

An endothermic animal can maintain a higher metabolic rate for a longer period due to

(A) increased ATP production from glycolysis
(B) a higher basal metabolic rate
(C) more efficient respiration
(D) greater retention of heat
(E) constant ATP production

QUESTION 7

What process utilizes the majority of energy produced by animals?

(A) movement
(B) metabolism
(C) homeostasis
(D) food processing
(E) vascular transport

QUESTION 8

The four mechanisms of heat transfer between an organism and its environment do NOT include

(A) radiation
(B) evaporation
(C) convection
(D) transpiration
(E) conduction

ANSWERS

1. D
2. B
3. A
4. C
5. E
6. B
7. C
8. D

ANIMALS
Digestion

IT'S ALIMENTARY

TOTALLY TUBULAR

Animals eat to fuel their bodies with chemical energy, organic molecules, and essential nutrients their bodies can't produce. Mammals have digestive tubes (**alimentary canals**) to process food.

FROM FOOD TO FECES

Food is broken down and processed in four stages: ingestion, digestion, absorption, and elimination.

Ingestion, or chewing and swallowing, involves the mouth, salivary glands, pharynx, and esophagus.

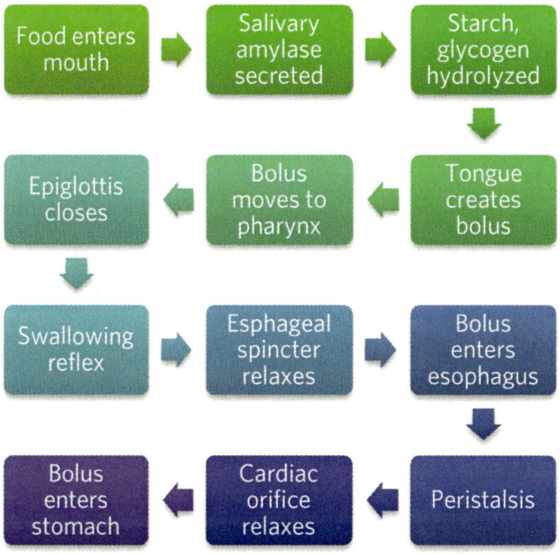

Digestion, or the enzymatic breakdown of food, starts with amylase in the oral cavity and continues in the stomach and small intestine.

THE STOMACH

In the stomach, pepsin digests proteins in the presence of strong acids and constant churning.

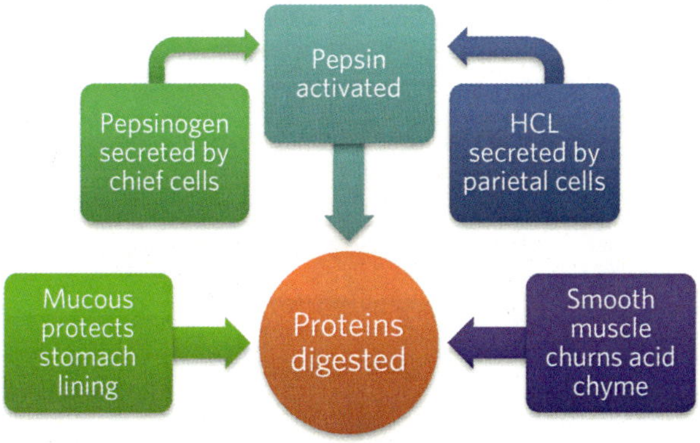

TWISTED TRACTS

DIGESTIVE SUPERHERO

The majority of digestion and nutrient absorbtion happens in the small intestine.

Digestion in the **small intestine** involves several glands secreting enzymes into the **duodenum** (the first 25 cm of the small intestine).

Gland/Structure	*Enzyme*	*Compound*
Pancreas	Amylase	Polysaccharide
	Trypsin Chemotrypsin	Polypeptide
	Carboxypeptidase	Small polypeptide
	Nuclease	DNA, RNA
	Lipase	Fat droplet
Gall bladder	Bile salts	Fat globule
Brush border	Disaccharidase	Disaccharide
	Proteases (3)	Small peptide
	Nucleotidase	Nucleotide
	Nucleosidase Phosphatase	Nucleoside

Absorption, or the transfer of hydrolyzed nutrients to the bloodstream, occurs in the small intestine. The **large intestine** reabsorbs water.

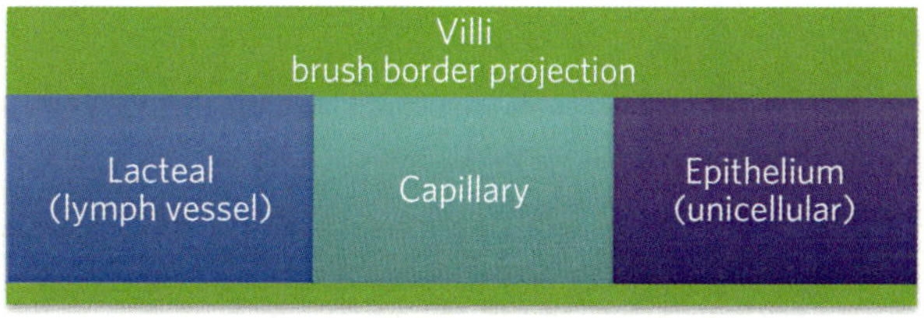

Elimination, or the removal of waste from the body, happens at the end of the colon (**rectum**), where two sphincters control the passage of feces through the **anus**.

HORMONAL CONTROL OF DIGESTION

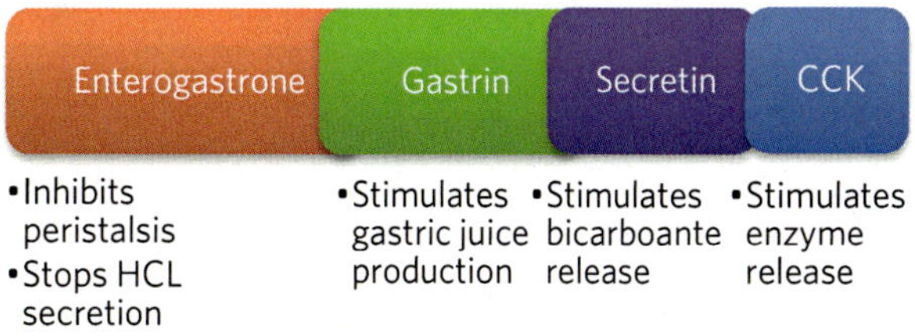

CRAM QUIZ
Digestion

QUESTION 1

What structures increase the surface area of the small intestines of earthworms, grasshoppers, and humans, respectively?

(A) dorsal ceca, crop, brush border
(B) food vacuoles, typhlosole, villi
(C) mesoglea, gastrodermis, gastric glands
(D) typhlosole, ceca, villi
(E) infoldings, several guts, multiple tracts

QUESTION 2

Which of the following substances are mixed in the duodenum by peristalsis?

(A) acid chyme, pancreatic juice, hydrolytic enzymes
(B) acid chyme and intestinal juices
(C) acid chyme and juices from the pancreas, gallbladder, and small intestine
(D) acid chyme and enzymes from the pancreas and gallbladder
(E) gastric juices

QUESTION 3

How does the endocrine system function in blood sugar regulation?

(A) Insulin inhibits glucose absorption when blood sugar drops.
(B) Glucagon stimulates the absorption of glucose by the body.
(C) Insulin encourages glucose storage when blood sugar drops.
(D) Glucagon stimulates glycogen breakdown in the liver.
(E) Insulin raises blood sugar, and glucagon lowers it.

QUESTION 4

Fat digestion and absorption occurs in what order? 1) Chylomicrons enter lacteals via exocytosis. 2) Bile salts emulsify fats. 3) Micelles diffuse into epitheleum. 4) Lipase forms micelles. 5) Micelles recombine. 6) Lymph distributes fatty substances.

(A) 2, 4, 3, 5, 1, 6
(B) 1, 4, 3, 5, 2, 6
(C) 2, 3, 5, 1, 4, 6
(D) 3, 5, 1, 2, 4, 6
(E) 2, 4, 1, 3, 5, 6

QUESTION 5

What functions do the crop and gizzard, respectively, perform in digestion?

(A) moistening and storage
(B) storage and nutrient absorption
(C) digestion and pulverization
(D) pulverization and nutrient absorption
(E) storage and pulverization

QUESTION 6

Why does the hepatic portal vein leave the small intestine and travel to the liver before entering other tissues?

(A) The liver's nutrient requirements are higher.
(B) The liver stores glucose.
(C) The liver processes many organic molecules.
(D) The liver prepares the blood to enter the heart.
(E) The liver synthesizes proteins

QUESTION 7

How do the digestive tracts of carnivores and herbivores differ?

(A) The large stomachs of carnivores encourage protein digestion.
(B) The small intestine is longer in herbivores.
(C) The cecum may promote fermentation in herbivores.
(D) Herbivores have a four-chambered stomach to enrich the food they eat before digestion.
(E) Carnivores have extensive cecums filled with bacteria that aid in digestion.

QUESTION 8

In hydras, which structures of the inner epidermis function in extracellular and intracellular digestion, respectively?

(A) mesoglea and gland cells
(B) gland cells and nutritive muscular cells
(C) food vacuoles and mesoglea
(D) gastrodermis and mesoglea
(E) nutritive muscular cells and gastrodermis

ANSWERS

1. D
2. C
3. D
4. A
5. E
6. C
7. C
8. B

ANIMALS
Circulation and Respiration

A WILD RIDE

FUNDAMENTAL FLOW

Animals transport gases and nutrients through a system of vessels that circulate through the body.

HAVE A HEART

The heart pumps blood through the circulatory system by contracting (**systole**) and relaxing (**diastole**). The **semilunar** and **atrioventricular** valves prevent backflow, and the **sinoatrial node** sets the heart rate by coordinating muscular contraction.

SMALL TO SMALLER

Vertebrate vessels carrying oxygenated or deoxygenated* blood get progressively smaller to facilitate chemical exchange with interstitial fluid.

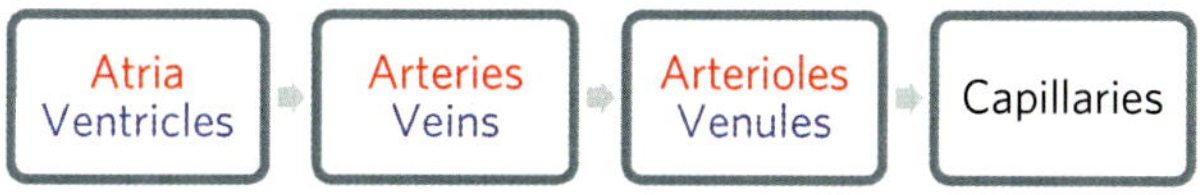

**See below for exceptions to this rule.*

TWICE TRANSPORTED

The mammalian cardiovascular system is divided into two circuits—**systemic** (to the body) and pulmonary (to the lungs)—that function simultaneously.

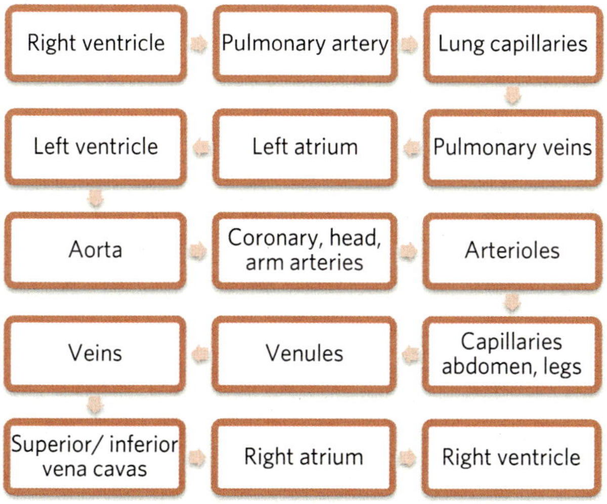

UNDER PRESSURE: THE LYMPHATIC SYSTEM

1. Blood exerts hydrostatic pressure against vessels.
2. Narrow openings of arterioles increase pressure inside arteries (**peripheral resistance**).
3. Pressure is high even when the heart relaxes.
4. Fluid flows from capillaries to the interstitial fluid.
5. Overflow (**lymph**) returns to blood after being filtered in the **lymph nodes**.

COUNTLESS CIRCUITS

THE BASICS OF BLOOD

Blood is mostly water containing dissolved ions, proteins, nutrients, wastes, gases, and cells.

Component	Function
Electrolytes	Buffering, membrane permeability
Albumin	Buffering
Fibrogen	Blood clotting
Immunoglobulins	Defense
Erythrocytes	Gas transport
Leukocytes	Defense
Platelets	Blood clotting

BETTER BREATHING

Mammals get the oxygen they need in cellular respiration from air. The lungs exchange carbon dioxide for oxygen by changing their internal pressure relative to the pressure outside (**negative pressure breathing**).

IN AND OUT

The **medulla oblongata** (with help from the **pons**) sets breathing rhythm as sensors in the **aorta** and **carotid arteries** (and medulla) monitor blood pH and oxygen levels.

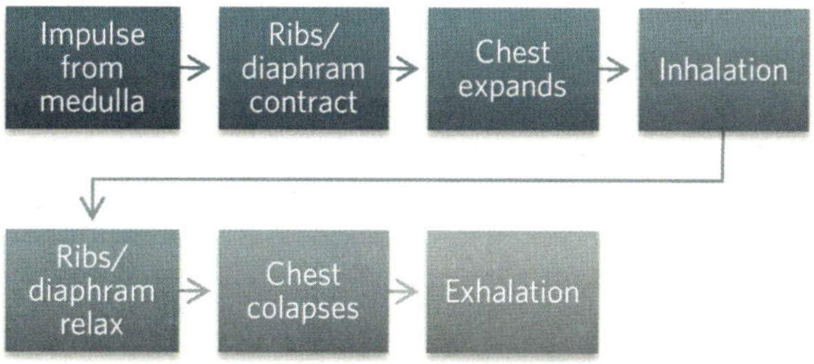

FOLLOW THE AIR FLOW

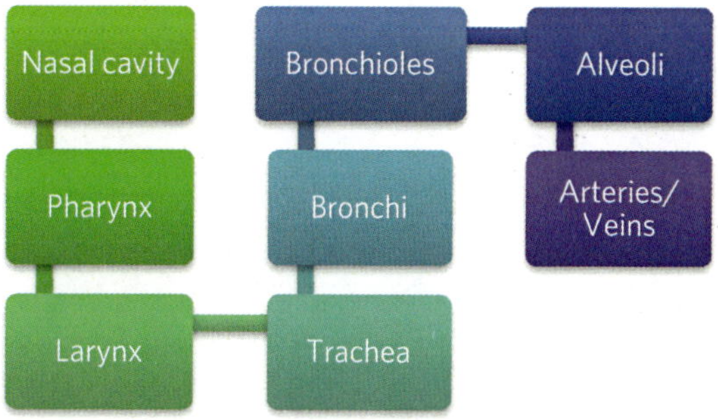

CRAM QUIZ
Circulation and Respiration

QUESTION 1

Which of the following is an advantage of closed circulation over open circulation?

(A) requires less energy to build
(B) structural support from a hydrostatic skeleton
(C) requires less energy to maintain
(D) transports more oxygen to cells
(E) blood circulates through the body twice

QUESTION 2

The spike in an ECG represents what part of a normal mammalian heart's cycle?

(A) ventricular contraction
(B) Purkinje fibers stimulation
(C) impulse from SA node
(D) signal delay at AV node
(E) atrial contraction

QUESTION 3

How do arteries and veins, respectively, keep blood flowing through the body?

(A) sphincters opening and closing; narrow valves
(B) smooth muscle contraction; skeletal muscle contraction
(C) closer to heart; closer to lungs
(D) peripheral resistance; smooth muscle contraction
(E) thick, elastic vessel walls; skeletal muscle contraction

QUESTION 4

What are the steps, in order, of blood clotting? 1) Plug forms. 2) Fibrogen converts to fibrin. 3) Platelets bind to collagen. 4) Prothromibin converts to thrombin. 5) Fibrin forms clot. 6) Platelets make other platelets sticky. 7) Endothelium damaged.

(A) 7, 3, 6, 1, 4, 2, 5
(B) 7, 1, 4, 6, 3, 2, 5
(C) 7, 2, 4, 3, 5, 6, 1
(D) 7, 3, 1, 4, 2, 6, 5
(E) 7, 2, 3, 5, 6, 1, 4

QUESTION 5

What is a disadvantage of having air as opposed to water as a respiratory medium?

(A) Gases diffuse more slowly in air.
(B) The respiratory surface loses water quickly.
(C) It requires more ventilation.
(D) Ventilation requires more energy.
(E) The respiratory surface is small.

QUESTION 6

What are the differences in partial pressures of oxygen and carbon dioxide in the cells (except lung tissue)?

(A) Carbon dioxide is twice as high as oxygen.
(B) Oxygen is slightly higher than carbon dioxide.
(C) Carbon dioxide is slightly higher than oxygen.
(D) Oxygen is very low.
(E) Carbon dioxide is extremely high.

QUESTION 7

What happens during the Bohr shift?

(A) Hemoglobin's affinity for oxygen increases.
(B) The amount of carbon dioxide in the blood drops.
(C) Cellular respiration decreases.
(D) Blood pH drops.
(E) Erythrocytes carry less hemoglobin.

QUESTION 8

Bicarbonate and hydrogen ions, respectively, are transported by which of the following to the lungs before being converted into carbon dioxide and water?

(A) hemoglobin; plasma
(B) erythrocytes; hemocyanin
(C) plasma; hemoglobin
(D) anhydrase; plasma
(E) leukocytes, erythrocytes

ANSWERS

1. D
2. A
3. E
4. A
5. B
6. C
7. D
8. C

ANIMALS
Immunity

DANGEROUS DEFENSES

TAKE NO PRISONERS

Animals attack bacteria, viruses, abnormal cells, and other pathogens in two ways: innate and acquired immunity.

FAST AND FAR-REACHING

Acquired immunity mounts a quick non-specific response to a variety of pathogens, without previous contact.

INTERNAL DEFENSES

The body has four internal defenses to fight pathogens that break its external barriers.

Phagocytes engulf and destroy pathogens.

Neutrophils	•Abundant in infected tissues; short lives
Macrophages	•Large, long lives, scarce
Eosinophils	•Fight parasites, secrete enzymes
Dendritic cells	•Engulf pathogens

Antimicrobial proteins in the blood are activated by pathogens. Together, they make up the **complement system** and burst microbial cells.

Natural killer cells attach to cancerous or virus-infected cells and lead to cell death (**apoptosis**).

The ***inflammatory response*** is triggered by cell damage or pathogens.

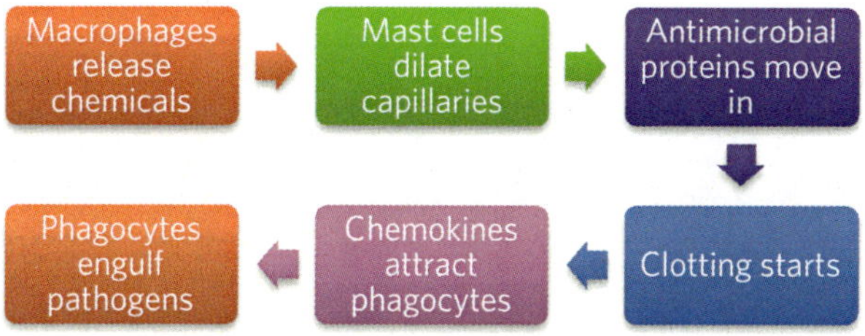

EXCELLENT ACQUISITION

PRECISE PROTECTION

Acquired immunity develops over time in response to specific pathogens by activating lymphocytes (**antibodies**).

PERFECT FIT

Antibodies respond to specific pathogens (antigens) by binding to small molecules on their surface called epitopes.

TWICE THE TROUBLE

The two types of acquired immunity, humoral and cell-mediated, differ in their defenses.

	Humoral	*Cell-mediated*
Location	Extracellular fluid	Infected tissues
Target	Pathogens	Infected, cancerous, foreign cells
Cells	Helper T, B, plasma, memory B	Helper T, cytotoxic T, memory cytotoxic T

HEFTY HELPERS

Helper T cells activate immunity by binding to MHC molecules on antigen-presenting cells.

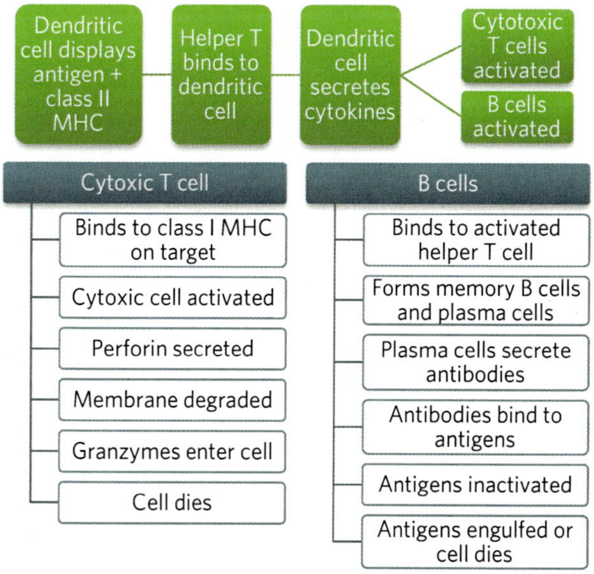

DEMIDEC

CRAM QUIZ

Immunity

QUESTION 1

What types of immune cells are present in the lymphatic system?

(A) mast cells, dendritic cells, lymphocytes
(B) neutrophils, eosinophils, lymphocytes
(C) macrophages, dendritic cells, lymphocytes
(D) monocytes, mast cells, antimicrobial proteins
(E) macrophages, neutrophils, dendritic cells

QUESTION 2

How do B and T cell receptors differ structurally?

(A) The B cell's polypeptide chains are linked by disulfide bridges.
(B) The T cell has two heavy and two light chains.
(C) The B has a disulfide bridge connecting its transmembrane region.
(D) The T cell's variable regions are Y-shaped.
(E) The B cell's chain have α and β conformations.

QUESTION 3

The secondary immune response is driven by what cloned immune cells?

(A) plasma cells
(B) effector cells
(C) cytotoxic T cells
(D) B cells
(E) memory cells

QUESTION 4

What is the role of class II MHC molecules in antigen presentation?

(A) They are displayed for cancerous or foreign cells.
(B) They are displayed by antigens and recognized by T cells.
(C) They create a MHC-carbohydrate-antigen complex.
(D) Class II MHC molecules bind to antigen fragments.
(E) They are always present on healthy cells.

QUESTION 5

The CD8 and CD4 surface proteins facilitate the binding of what immune cells, respectively?

(A) cytotoxic T cells; B cells
(B) cytotoxic T cells; helper T cells
(C) B cells; cytotoxic T cells
(D) helper T cells; B cells
(E) helper T cells

QUESTION 6

What is true of secreted antibodies called immunoglobulins?

(A) They belong to four classes.
(B) They have a transmembrane region.
(C) The tail region determines their functions.
(D) They are protected from cleavage.
(E) The IgE class is the most abundant.

QUESTION 7

How do antibodies function in antigen removal?

(A) They lead to phagocytosis of infected cells.
(B) They block a virus' ability to replicate.
(C) They form chains of microbes.
(D) They keep antigens in solution.
(E) They activate the complement system.

QUESTION 8

HIV enters helper T cells, brain cells, and macrophages by binding to what surface proteins?

(A) CD4 and chemokine receptor
(B) CD8 and a coreceptor
(C) CD8 and class I MHC
(D) CD4 and class I MHC
(E) CD4 and class II MHC

ANSWERS

1. C
2. A
3. E
4. D
5. B
6. C
7. E
8. A

ANIMALS
Homeostasis

ROCK STEADY

CAN'T WIN FOR LOSING

To maintain a consistent internal environment, animals balance gaining and losing heat and water in a process called homeostasis.

BODIES IN MOTION

Cellular processes change the makeup of body fluids and produce toxic wastes. Animals fight these changes by excreting wastes from the body.

Filtration
- Blood pressures forces fluid into excretory tubules

Reabsorption
- Needed solutes are actively pumped back to capillaries

Secretion
- Wastes and excess ions are actively pumped into excretory tubules

Excretion
- Processed fluid (filtrate) leaves the body as urine

FOLLOW THE FILTRATE

Excretion takes place in the kidneys' functional unit—the **nephron**—a tube surrounded by capillaries with a bulbous end (**glomerulus**).

Kidney region	*Part of nephron*	*What happens*
Cortex	Proximal tubule	Secretes HCO_3-, NaCl, H_2O, K+, solute; absorbs H+, NH_3
Outer medulla	Descending loop of Henle	Reabsorbs H_2O
Inner medulla	Ascending loop of Henle	Secretes NaCl by diffusion
Outer medulla	Ascending loop of Henle	Secretes NaCl by active transport
Cortex	Distal tubule	Reabsorbs K+, H+; secretes NaCl, H_2O, HCO_3-
Outer medulla	Collecting duct	Secretes NaCl
Inner medulla	Collecting duct	Secretes urea, H_2O

REAL REGULATORS

COUNTERCURRENT CONCENTRATION

As filtrate travels through the nephron's winding tubules, water diffuses into the kidney and urea becomes concentrated in the collecting duct.

Water diffuses into kidneys, osmolarity increases → NaCl transported into kidneys, osmolarity decreases → Water diffuses into kidneys, urea's osmolarity increases

NEGATIVE FEEDBACK

Several hormones control the solute concentration and volume of urine excreted by the body.

Osmolarity
- Hypothalamus detects osmolarity increase
- ADH secreted
- Collected duct permeability increases; Thirst stimulated

Blood pressure, volume
- JGA detects low volume/pressure
- Renin produced; Angiotensinogen converts to angiotensin II
- Arteries constrict; Adrenals secrete alderosterone, increasing reabsorption

TOO HOT TO HANDLE

Animals maintain a constant internal body temperature through the process of thermoregulation with the help of several adaptations.

Circulation
- Vasodilation
- Vasoconstriction
- Countercurrent heat exchange

Evaporation
- Breathing, panting
- Sweating

Insulation
- Feathers
- Hair, fur
- Fat

Behavior
- Hibernation
- Migration

Metabolism
- Shivering, movement
- Nonshivering thermogenesis
- Brown fat

DemiDec

CRAM QUIZ
Homeostasis

QUESTION 1

Aquatic species tend to excrete ammonia because

(A) most waste is excreted by the skin
(B) it is soluble in water
(C) their gills tolerate high concentrations of ammonia
(D) their kidneys tolerate high concentrations of ammonia
(E) urea is more toxic

QUESTION 2

Insects and reptiles excrete uric acid

(A) through nephridiopores on the body's surface
(B) to balance the influx of water to the body
(C) because it is soluble in water
(D) as dry matter
(E) after pumping water back into the body at the rectum

QUESTION 3

In humans, most nephrons are confined to what part(s) of the kidney?

(A) cortex
(B) inner medulla
(C) outer medulla
(D) outer and inner medulla and cortex
(E) outer and inner medulla

QUESTION 4

The substances that are actively transported from the nephron to the kidneys during excretion are

(A) HCO_3- NaCl, and K+
(B) H+, NH_3, and solutes
(C) urea, solutes, and NaCl
(D) K+, solutes, and HCO_3-
(E) HCO_3-, NaCl, and solutes

QUESTION 5

The difference in the osmolarity of the interstitial fluid of the proximal and distal tubules of the nephron is

(A) the osmolarity in the distal tubule is slightly higher
(B) the osmolarity in the distal tubule is twice as great
(C) the osmolarity in the proximal tubule is slightly higher
(D) the osmolarity in the proximal tubule is half as much
(E) very small

QUESTION 6

A rise in blood pressure and volume causes

(A) a decrease in renin production
(B) secretion of angiotensinogen
(C) conversion of angiotensin to angiotensiogen
(D) adrenal gland stimulation
(E) stimulation of RAAS

QUESTION 7

What is the function of ADH when a person is severely dehydrated?

(A) maintain osmolarity
(B) none
(C) lower solute concentration
(D) maintain blood pressure and volume
(E) stimulate water reabsorption

QUESTION 8

The atrial walls release ANF when

(A) RAAS levels are low
(B) osmolarity increases
(C) blood pressure and volume increases
(D) the salt concentration is high
(E) water should be reabsorbed

ANSWERS

1. B
2. D
3. A
4. E
5. C
6. A
7. B
8. C

ANIMALS
Hormones

CREATIVE COMMUNICAITON

FAST AND FLEXIBLE

Animals regulate body functions by secreting hormones into the bloodstream that then act on specific target cells.

GLANDS OR NEURONS?

The nervous and endocrine systems regulate hormone secretion both separately and in conjunction.

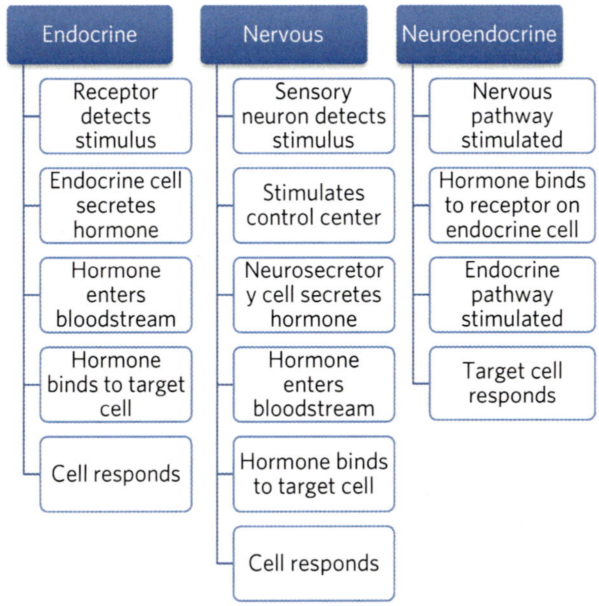

A CLOSER LOOK: GLANDS

- The **pancreas** controls blood glucose levels by releasing two antagonistic hormones: **insulin** and **glucagon**

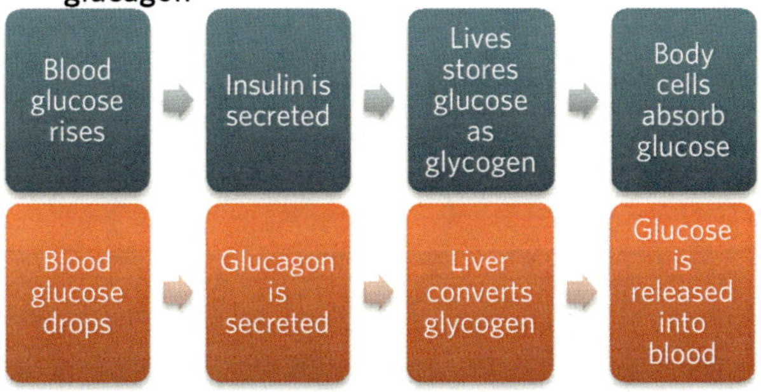

- The **pineal gland** secretes melatonin to regulate circadian rhythms and daily or seasonal behaviors
- The **gonads** regulate reproduction and the development of secondary sex characteristics

Testes	Ovary
•Androgens	•Estrogen, progestin

GREAT GLANDS

THYROID/PARATHYROID

- The parathyroid gland raises blood calcium with PTH
- The thyroid gland lowers blood calcium with **calcitonin** (and regulates metabolism with TSH)

HYPOTHALAMUS/PITUITARY GLAND

The hypothalamus of the brain regulates the anterior and posterior pituitary glands with tropic hormones, which in turn control reproduction, metabolism, and homeostasis.

Pituitary Gland	Hormone	Target
Posterior	ADH	Kidneys
	Oxytocin	Mammary glands, uterus
Anterior	FSH/LH	Testes/ovaries
	Prolactin	Mammary glands
	Growth hormone	Liver, bones
	TSH	Thyroid
	ACTH	Adrenal cortex
	MSH	Melanocytes
	Endorphin	Brain

ADRENAL MEDULLA/CORTEX

Hormones secreted by the adrenal gland's two regions respond to stress in the short and long terms.

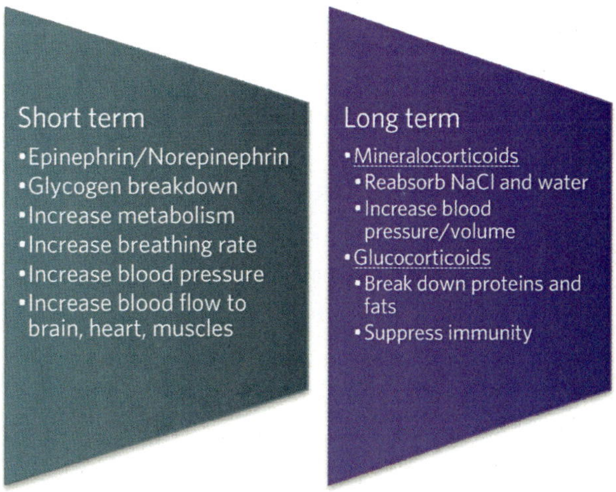

CRAM QUIZ
Hormones

QUESTION 1

A single hormone can elicit different responses in target cells by

(A) changing conformation
(B) binding to coreceptors
(C) initiating different signal transduction pathways
(D) being secreted by different glands
(E) functioning as a nerve and endocrine signal

QUESTION 2

The difference between paracrine and hormone signaling is

(A) hormones are faster-acting
(B) paracine signaling is local
(C) endocrine signaling elicits different responses
(D) paracrine signaling is initiated by control centers
(E) hormones don't bind to receptors

QUESTION 3

How does nitric oxide function in paracrine signaling?

(A) By contracting smooth muscle and dilating vessels
(B) similarly to the drug Viagra
(C) by being readily available
(D) as a neurotransmitter
(E) by contracting muscles during labor

QUESTION 4

How do water and lipid-soluble hormones differ?

(A) Lipid-soluble hormones travel across the plasma membrane.
(B) Lipid-soluble hormones initiate signal transduction pathways.
(C) Water-soluble hormones bind to receptors inside the nucleus.
(D) Water-soluble hormones diffuse across the secretory cells' membranes.
(E) Lipid-soluble membranes are secreted via exocytosis.

QUESTION 5

The hormones responsible for negative feedback on the hypothalamus and anterior pituitary are

(A) TSH, TRH
(B) thyroid secreted hormones
(C) thyroxine
(D) triiodothyronine
(E) TRH

QUESTION 6

Blood calcium levels decrease when

(A) calcitonin is released
(B) bones demineralize
(C) the kidneys reabsorb calcium
(D) vitamin D is activated
(E) the parathyroid gland is stimulated

QUESTION 7

In type I diabetes mellitus,

(A) insulin is broken down
(B) blood glucose drops
(C) the liver no longer reacts to insulin
(D) the pancreas can't make glucagon
(E) the immune system destroys beta cells

QUESTION 8

Pupa formation in insects occurs when

(A) ecdysone is released
(B) brain hormone stimulates ecdysone
(C) juvenile hormone concentrations fall
(D) the prothoracic gland inhibits ecdysone production
(E) juvenile hormone is released

ANSWERS

1. C
2. B
3. D
4. A
5. B
6. A
7. E
8. C

ANIMALS
Reproduction and Development

LET'S TALK ABOUT SEX

MAKIN' GAMETES

Many animals have gonads that produce gametes, along with other structures that facilitate reproduction.

HUMAN GENESIS

In humans, oogenesis and spermatogenesis produce eggs and sperm, respectively.

Oogenesis
- Oogonium in ovary (2n)
- Mitosis
- Primary oocyte (2n)
- FSH stimulates follicle
- Meiosis I and part of meiosis II
- Secondary oocyte, first polar body (n)
- Ovulation
- Sperm penetration
- Meiois II completed

Spermatogenesis
- Spermatogonium in testis (2n)
- Mitosis and meiosis start
- Primary spermatocyte (2n)
- Meiosis I ends
- Secondary spermatocyte (n)
- Meiosis II
- Early spermatids (n)
- Differentiation
- Sperm cell

GO WITH THE FLOW

Several hormones regulate the timing of the uterine cycle and the ovarian cycle.

- Follicle grows / Menstruation / Estrogen rises
- Follicle matures / Endometrium grows / LH, FSH spike
- Ovulation / Endometrium grows
- Corpus luteum forms / Endometrium grows / Estrogen, progesterone rise
- Corpus luteum disintegrates / Endometrium grows

SUCCESS!

MAKIN' BABIES

Mammalian fertilization and the stages of embryonic development occur inside the womb, unlike the external fertilization of some marine animals.

STEP 1: FERTILZE

Only one sperm enters an egg thanks to the cortical reaction.

Sperm enters follicle → Binds to zona pellucida → Hydrolytic enzymes are released → Zona pellucida breaks down → Binds to plasma membrane → Membranes fuse → Sperm nucleus enters egg → Cortical granules fuse to membrane → Zona pellucida hardens

STEP 2: GROW

Fertilization is followed by implantation of the zygote in the endometrial lining of the uterus.

Sperm and egg nuclei fuse → Zygote forms → Cleavage begins → Zygote travels to uterus via oviduct → Zygote remains free in uterus → Blastocyst forms → Blastocyst implants in endometrium

STEP 3: DEVELOP

After implantation, the **epiblast** (upper cell layer) of the blastocyst undergoes cell differentiation (**gastrulation**) followed by organ development (**organogenesis**).

Epiblast →
- Mesoderm
  - Skeletomuscular, circulatory, lympathic, reproductive, and excretory systems
- Ectoderm
  - Skin, epidermis, and nervous system
- Endoderm
  - Liver, pancreas, thyroid, para-thyroid, and lining of digestive and respiratory systems

CRAM QUIZ
Reproduction and Development

QUESTION 1

Which of the following substances does the prostate gland contribute to seminal fluid?

(A) fructose, ascorbic acid, and prostaglandins
(A) anticoagulants and citrate
(B) mucus and sperm
(C) mucus, coagulant, and citrate
(D) fructose, ascorbic acid, and prostaglandins

QUESTION 2

Human chorionic gonadotropin, which stimulates estrogen and progesterone secretion, is secreted by the

(A) ovaries
(B) corpus luteum
(C) embryo
(D) anterior pituitary
(E) uterus

QUESTION 3

The hormones that regulate lactation include

(A) progesterone and oxytocin
(B) prolactin and estrogen
(C) prolactin and progesterone
(D) prolactin and oxytocin
(E) oxytocin and estrogen

QUESTION 4

Why doesn't the mother's immune system reject the developing fetus?

(A) The trophoblast may inhibit the immune response.
(B) The placenta suppresses the mother's B cells.
(C) Placental enzymes break down the mother's T cells.
(D) The trophoblast causes T cell apoptosis.
(E) The placenta produces an enzyme to synthesize tryptophan, which inhibit T cells.

QUESTION 5

A major difference between early bird and mammalian development is

(A) gastrulation
(B) organogenesis
(C) fertilization
(D) invagination
(E) cleavage

QUESTION 6

The notocord of a developing frog embryo induces the formation of what structure(s)?

(A) somites
(B) neural plate
(C) archenteron
(D) mesoderm
(E) coelom

QUESTION 7

What happens if the gray crescent of a frog embryo is limited to one side of the embryo?

(A) The embryo does not differentiate.
(B) Cleavage stops.
(C) Dorsal structures on that side don't develop.
(D) The cells remain totipotent.
(E) Development stops after gastrulation.

QUESTION 8

The process by which cells become longer and narrower is called

(A) convergent extension
(B) induction
(C) hinging
(D) invagination
(E) involution

ANSWERS

1. B
2. C
3. D
4. A
5. E
6. B
7. C
8. A

ANIMALS
Nervous System

VERY IMPULSIVE

ACTION DRIVEN

Neurons send signals from the central nervous system (the brain and spinal cord) to the peripheral nervous system by changing their membrane potentials.

HAVING THE POTENTIAL

Neurons generate impulses through action potentials.

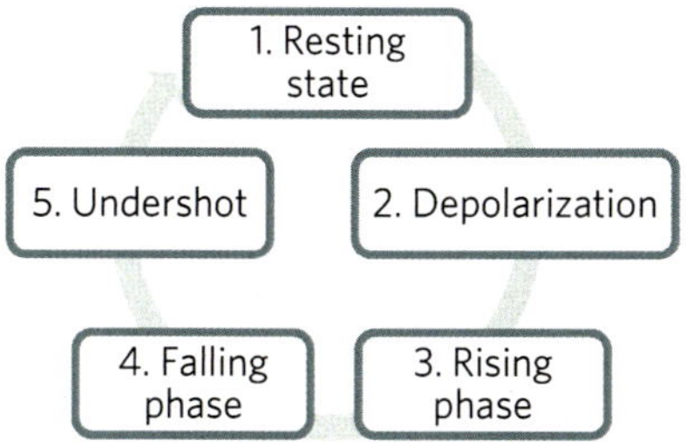

1. Sodium and potassium channels are closed.
2. Sodium channels open. Sodium enters the cell. The cell reaches threshold.
3. The inside of membrane is positively charged compared to outside.
4. Sodium channels close. Potassium channels open. Potassium leaves the cell.
5. The inside of the cell is negative compared to the outside. Potassium channels close. Sodium channels begin to open.

SEPARATE BUT EQUAL

Neurons may propagate impulses by releasing **neurotransmitters**, which bind to ion channels on the **postsynaptic** membrane and generate EPSPs or IPSPs.

Added together (**summation**), EPSPs generate action potentials; IPSPs inhibit action potentials.

Neurotransmitter	*Inhibitory*	*Excitatory*
Acetylcholine	✓	✓
Dopamine		✓
GABA	✓	
Glycine	✓	
Glutamate		✓
Serotonin	✓	
Norepinephrine	✓	✓
Aspartate		✓

SENSIBLE STRUCTURES

THE FAB FIVE

Specialized structures use the five senses (hearing, seeing, smelling, tasting, and touching) to detect changes in the environment and relay that information to the brain.

SAY WHAT?

The human ear collects sound and funnels it to specialized neurons that convert sound to impulses.

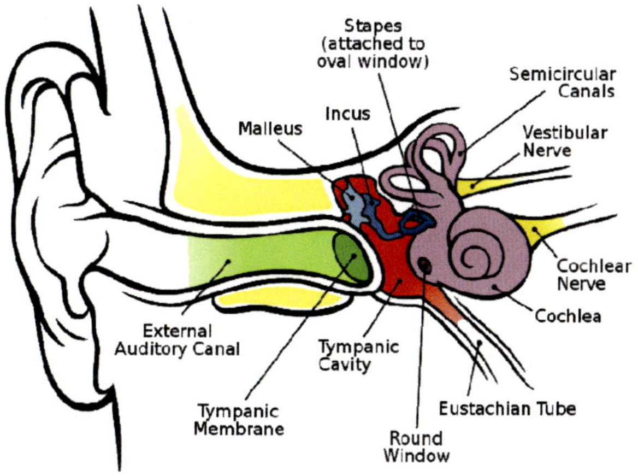

EYES WIDE OPEN

Human eyes contain visual pigments that transmit light energy to neurons, creating visual images with depth, contrast, and color.

Eye cell	*Function*
Ganglion	Synapse bipolar cell; transmits impulses
Horizontal	Integrates sensory information
Amacrine	Integrates sensory information
Bipolar	Synapse rods and cones; excited or inhibited
Rod	Sensitive to light
Cone	Distinguishes colors

TASTING YOUR SMELLS

Taste and smell are separate pathways that interact.

Taste
- Receptor binding, signal transduction, depolarization, neurotransmitter release

Smell
- Receptor binding, depolarization

CRAM QUIZ
Nervous System

QUESTION 1

What is the function of the interneurons in a reflex arc?

(A) detect stretching in muscles
(B) send signals to muscles
(C) inhibit motor neurons
(D) transmit information to the spinal cord
(E) stretch muscle cells

QUESTION 2

Which support (glia) cells induce the formation of the blood-brain barrier?

(A) Schwann cells
(B) oligodendrocytes
(C) radial glia
(D) astrocytes
(E) myelin

QUESTION 3

Which of the following statements correctly describes the speed of action potential transmission?

(A) Wider axons conduct impulses faster.
(B) Greater resistance speeds up transmission.
(C) Myelin sheaths increase speed by bringing closer regions of the neuron to threshold.
(D) It slows down when channels are concentrated at gaps in the myelin sheath.
(E) It is faster at the postsynaptic membrane.

QUESTION 4

The parasympathetic, sympathetic, and enteric divisions, respectively, of the autonomic nervous system function in

(A) peristalsis, erection, and pupil dilation
(B) ejaculation, decreasing heart rate, and salivation
(C) liver stimulation, stomach stimulation, and urination
(D) gallbladder stimulation, intestinal stimulation, and erection
(E) urination, raising heart rate, and peristalsis

QUESTION 5

Long-term potentiation in the brain involves

(A) potassium ions and AMPA receptors
(B) glutamate and NMDA receptors
(C) GABA and calcium ions
(D) nitric oxide and glutamate
(E) calcium ions and glycine

QUESTION 6

What transmits sound waves in the inner ear?

(A) tympanic canal
(B) basilar membrane
(C) perilymph
(D) vestibular canal
(E) stapes

QUESTION 7

During accommodation in the human eye,

(A) ciliary muscles relax
(B) the choroid moves away from lens
(C) suspensory muscles pull the lens
(D) ciliary muscles contract
(E) the lens flattens

QUESTION 8

What happens to rod cells when the human eye sees light?

(A) Sodium channels in the plasma membrane close.
(B) The cells are depolarized.
(C) The cells secrete glutamate.
(D) The cells depolarize bipolar cells.
(E) The rhodopsin in the cell is inactive.

ANSWERS

1. C
2. D
3. A
4. E
5. B
6. C
7. D
8. A

TEST-TAKING STRATEGY
Introducing the Exam

HIGHER LEARNING

YOU'VE COME A LONG WAY

The AP Biology course is far from easy—it covers all the material you would learn in an introductory biology course in college. This means you probably know as much about biology as many college students. So remember, you're already way ahead of the pack.

The field of biology is changing rapidly. It's impossible to know every detail. The purpose of AP Biology is to teach not only facts and concepts, but also critical thinking and the process of science. Your laboratory experience and introduction to the major themes in biology provide an excellent foundation for classroom tests and the AP Biology exam.

STUDYING SUCCESSFULLY

In the classroom and laboratory you've learned (or will learn) everything from fungal reproduction to membrane structure. This overwhelming store of information is more manageable if you divide it into five categories. Keep these categories in mind while studying to avoid confusion.

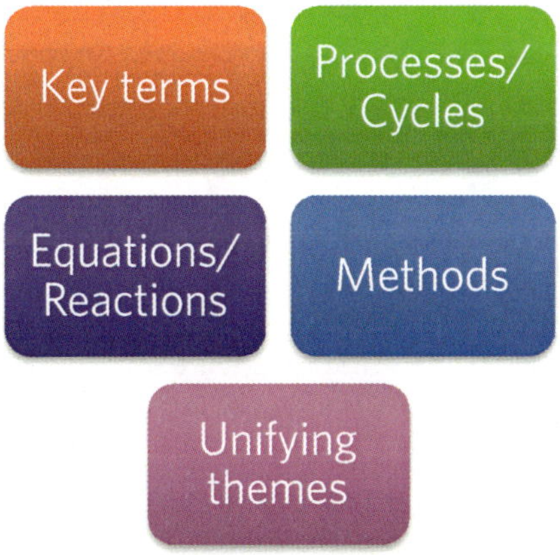

Key terms are essential words or phrases that are usually bolded or italicized in your textbook (and in this book, too!). Knowing these terms means you have a good handle on important concepts.

Processes and cycles like digestion and mitosis are usually outlined by helpful diagrams or illustrations.

Equations and reactions like photosynthesis or the population growth model should be memorized.

Methods are the scientific processes that led to major discoveries. Mendel's genetic experiments with pea plants are an example.

Unifying themes are underlying concepts in biology.

GETTIN' DOWN TO BUSINESS

THE AP BIOLOGY TEST

After nine months of hard work, it's finally here! Preparing for the AP Biology test is a cinch if you know how it's structured and what to study.

The AP Biology exam is split into two parts.

- Part I includes **100 multiple-choice questions** that make up 60% of your final score
- Part II includes **four essay questions** that make up 40% of your final score

How many questions will you find on each topic? That's easy to calculate.

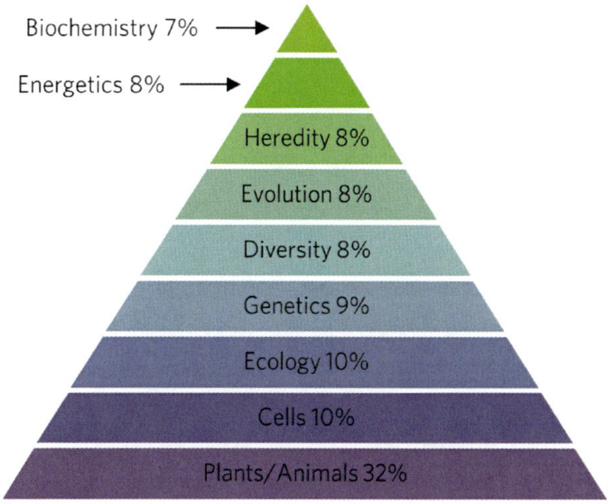

The AP exam will follow these percentages pretty closely. In other words, expect to see about 10 multiple-choice questions about ecology, and so on.

After the multiple choice section, you'll face four essays:

- Two essay questions focus on organisms and populations
- One focuses on molecules and cells
- One focuses on heredity and evolution

If you'd like a more detailed topic outline, visit the College Board's website (collegeboard.com) for the AP Biology exam.

Here are a few important things to keep in mind about the test:

- Don't be surprised by unfamiliar terminology—the test writers don't expect you to know everything
- Understand how the test is graded because this will affect how you answer test questions (see the next page for more on this)

TEST-TAKING STRATEGY
Introducing the Exam

SO MANY CHOICES

MULTIPLE CHOICE

The AP Biology test includes 100 multiple-choice questions that make up 60% of your final score. Multiple-choice questions seem easy on the surface—I mean, the answer's right there! But there are lots of ways multiple-choice questions can trick or confuse hapless test-takers.

To avoid these traps, remember to:

- Read the question carefully
- Look for words like NOT, EXCEPT, FALSE, or TRUE
- Eliminate answers that are obviously wrong
- Circle the correct answer (in your test booklet) or make your best guess
- Fill in the circles on your answer sheet neatly
- Don't spend too much time on any one question

IMPORTANT NOTE!

Right answers = points

Blank answers = no points

Wrong answers = **no penalty** (as of 2011)

In other words, **you should ALWAYS guess**, even if you're clueless, on a multiple-choice question.

Still, if you have time, try to eliminate as many wrong answers as possible first.

MORE QUICK TIPS

- Bring several sharpened #2 pencils to the test
- Multiple-choice questions often cover specific, detailed information; studying key terms will help
- It's hard to cram details into your head at the last minute, so get studying as early as possible
- Focus on reading the question carefully; cover up the answers if it helps
- Remember you are looking for the BEST answer, so don't choose the first answer you think is correct; read all the choices
- Educated guessing (but not random guessing) will work to your advantage
- Look for verbal clues to help find the right answer; choices with conditional words like "probably," as opposed to absolute ones like "never," are more likely to be correct
- If you have time, review your answers after you're done

EXPRESS YOURSELF

ESSAYS

The AP Biology exam includes four essay questions that make up 40% of your final score. Essay questions are not as bad as they seem. These questions allow you branch out and let your creativity and voice shine. There are lots of ways to get credit for answering these questions, so let your pencil fly!

You've probably never encountered essay questions like the ones on the AP Biology test. Remember how the questions are graded to improve your score.

What's important
- Correct information
- Not contradicting yourself
- Labeled, titled diagrams
- Labeling answers
- Including basic information
- Answering the entire question
- Answering all questions

What's NOT important
- Incorrect information
- Formal essay structure
- Spelling, grammar, and general writing skills

Remember:
You don't get points off for incorrect information!

SUCCESSFUL STRATEGIES

Now that you know how the essays are graded, it's time to think about how to tackle the questions.

The College Board offers several tips to help students improve their essays.

1. An extra sheet of paper with the essay questions is provided. Use the insert to organize your thoughts, but remember to transfer ALL the important information to the answer sheet. Only the answer sheet is graded.
2. Read all the questions before you start writing and tackle the ones you know first.
3. Even if you give the wrong answer, you may receive credit for correct thinking, so always SHOW YOUR WORK.
4. If you write a lot of wrong information hoping something is right, the grader may take off points.
5. Neat handwriting isn't as important as well-organized material. If you make a mistake, don't waste time by erasing—just cross it out.

CRUNCH KIT
AP Biology in Two Pages (Page 1)

CARBON
- Carbon binds covalently to at most four other atoms with branching, ring, and linear structures
- Carbon, hydrogen, oxygen, and carbon make up most organic molecules
- Organic compounds have functional groups that participate in reactions

WATER
- Water is shaped like a V with slightly charged ends that make it a polar molecule
- Water forms hydrogen bonds with other water molecules and different molecules
- Water supports life through its unique properties: cohesion, adhesion, temperature moderation, and floating ice
- Water dissociates into hydrogen and hydroxide ions to create acidic or basic solutions
- The pH scale measures hydroxide ion concentration
- Hypertonic solutions have a higher solute concentration than hypotonic solutions

CELL TYPES
- Eukaryotic cells have nuclei and are compartmentalized to separate cellular processes
- Prokaryotic cells have no nuclear envelope to separate DNA from the rest of the cell
- Animal cells have lysosomes and centrioles
- Plant cells have a cell wall, chloroplasts, central vacuole, tonoplast, and plasmodesmata

CELL CYCLE
- Internal and external signals can cause checkpoints to stop or start cell division
- Mitosis produces diploid daughter cells with the same number of chromosomes as the parent
- Meiosis produces haploid daughter cells with half the number of chromosomes as the parent
- When division goes awry, the cell can grow uncontrollably

LIFE CYCLES
- Prokaryotes reproduce by binary fission
- Sexual reproduction combines DNA from two parents to produce genetically different offspring
- Animals have diploid multicellular organisms, haploid gametes, and diploid zygotes
- Plants and some algae produce haploid multicellular gametophytes and diploid multicellular sporophytes
- Plant and algae life cycles are called alternation of generations
- Fungi and protists have haploid multicellular organisms and diploid zygotes
- Sexual reproduction creates genetic diversity
- Asexual reproduction produces more offspring

METABOLISM
- Energy coupling takes free energy from exergonic reactions and uses it to drive endergonic reactions
- In coupled reactions, the free energy change for the overall reaction is negative
- ATP hydrolysis fuels other reactions and powers cellular processes
- ATP is recycled by cellular processes that yield energy
- Feedback mechanisms control cellular processes, often by stimulating or inhibiting enzymes

GENETIC DIVERSITY
- Independent assortment of chromosomes during meiosis produces genetic diversity
- Crossing over during prophase I of meiosis I shuffles genes and creates genetic diversity
- X inactivation and genomic imprinting add to genetic diversity

DNA REPLICATION
- Begins at the origin of replication and expands laterally until two daughter strands of DNA form
- DNA polymerase adds nucleotides to the 3′ end of an elongating DNA molecule
- The leading strand grows continuously
- The lagging strand is synthesized in fragments

TRANSCRIPTION
- RNA synthesis or transcription occurs in the nucleus of eukaryotes and the cytosol of prokaryotes
- Transcription occurs in three phases: initiation, elongation, and termination
- After transcription, the pre-mRNA is processed by removing introns and splicing together exons

TRANSLATION
- Protein synthesis or translation takes place on ribosomes
- Eukaryotes need transcription factors to start transcription
- Each mRNA has a codon or group of three bases that complements the anticodon on a tRNA molecule
- Every tRNA molecule binds a specific amino acid
- After translation, a protein may be modified and targeted for the endoplasmic reticulum
- Protein folding occurs at four levels that create complex structures

EVOLUTION
- Some individuals will be better adapted to an environment and are more likely to reproduce
- Populations, not individuals, evolve
- Natural selection and adaptation change the genetic makeup of populations and may create new species
- Evolution can be observed in living populations and the fossil record

CRUNCH KIT
AP Biology in Two Pages (Page 2)

DIVERSITY
- The first cells evolved from droplets of self-replicating organic molecules surrounded by membranes
- Early prokaryotes absorbed other prokaryotes to form early eukaryotes
- Currently, organisms are divided into three superkingdoms, or domains: Bacteria, Archaea, Eurkarya
- The biomass of prokaryotes is greater than thato f eukaryotes
- Protists may be more closely related to animals or plants than to other protists
- Land plants evolved from green algae
- Many flowering plants and pollinators have close evolutionary ties
- Fungi evolved from aquatic protists
- Animals are characterized by body plans

PLANTS
- Plants are incredibly responsive to the environment, which is also called plasticity
- Plants grow throughout their lifetimes, which is called indeterminate growth
- Apical meristems in roots and shoots contain embryonic cells
- Primary growth in plants adds length
- Secondary growth adds width or girth to woody plants

PLANTS – TRANSPORT
- Roots take in water, nutrients, and gases from the soil
- Xylem sap transports nutrients to the leaves via bulk flow
- Transpiration from the leaves pulls xylem sap to the top of the plants
- Sugars travel down the plant as phloem sap via pressure flow
- Fungi in roots help plants fix nitrogen

PLANTS – REPRODUCTION
- Plants can reproduce sexually or asexually, depending on the conditions
- Barriers prevent self-fertilization and encourage genetic diversity
- Double fertilization produces ovules surrounded by ovaries
- Ovules develop into seeds and ovaries into fruit

PLANTS – RESPONSES
- Plants use signal transduction pathways to respond to stimuli
- Plant hormones coordinate activities such as growth
- Hormones typically don't function in isolation but in conjunction with other hormones

ANIMALS
- Animals have specialized cells organized into tissues, organs, and organ systems

ANIMALS, CONTINUED
- Some animals maintain a constant internal environment with homeostasis
- Local or long distance signals may coordinate cellular processes in animals

ANIMALS – SYSTEMS
- Animals have digestive systems that reflect what they eat
- Animals must consume carbon and essential amino acids, fatty acids, vitamins, and minerals
- Digestion is either intracellular or extracellular
- Digestion occurs in four stages: ingestion, digestion, absorption, and elimination
- Invertebrate, fish, amphibian, reptilian, and mammalian circulatory systems differ
- Mammalian circulation occurs in two circuits that operate at the same time
- The heart controls heart rate and stroke volume involuntarily
- Respiratory mediums and surfaces differ among animals
- Respiratory pigments transport oxygen and carbon dioxide
- The immune system provides specific and nonspecific defenses against pathogens
- Antibodies recognize a portion of an antigen called an epitope on a pathogen
- Lymphocytes are responsible for acquired immunity and may secrete antibodies
- Animals secrete different forms of nitrogenous waste
- The kidneys maintain water and salt balance in the blood
- The nervous and endocrine systems regulate physiology
- The stages in animal development are fertilization, cleavage, gastrulation, and organogenesis
- The nervous system communicates with the body via electrical impulses
- The nervous system is specialized by body region
- The brain is functionally divided into different regions

ECOLOGY
- Ecology combines many different disciplines and can be difficult to quantify
- Abiotic and biotic factors affect organisms and climate
- Species distribution is limited by the interaction between organisms and their environment
- Ecology studies organisms at the levels of populations, communities, and ecosystems
- Organic and inorganic nutrients cycle through ecosystems
- Humans are impacting ecosystems by altering nutrient cycling, reducing biodiversity, and destroying habitats
- Researchers in ecology are finding ways to reverse the negative impacts of human populations on the environment

CRUNCH KIT
List of Lists

TOP 15 ORGANELLES

Nuclear envelope	Lined by the nuclear lamina; surrounds chromosomes
Nucleolus	Synthesizes RNA; assembles ribosomes
Ribosomes	Carry out protein synthesis; made of RNA and protein
ER	Part of endomembrane system; ships proteins to Golgi; smooth or rough
Golgi apparatus	Ships transport vesicles; *cis* and *trans* faces
Lysosomes	Engulf and break down particles
Vacuoles	Pump out water; transport sap; store food
Mitochondria	Convoluted membranes that carry out cellular respiration
Chloroplasts	Convert light to chemical energy
Peroxisomes	Break down compounds by oxidation
Cytoskeleton	Network of dynamic fibers throughout cell; changes shape, movement, activities
Plasma membrane	Surrounds cell; transports materials; hydrophobic core and hydrophilic exterior
Plasmodesmata	Channels of cytoplasm; connect plant cells
Junctions	Connect animal cells; types: tight, desmosomes, and gap
Extracellular	Outside the plasma membrane; cell wall in plants; glycoprotein matrix in animals

KNOWING HEREDITY

Dominant trait	Trait that is dominant and physically visible in an organism
Recessive trait	Trait that is dominated and not physically visible in an organism
Genotype	Genetic makeup of an organism
Phenotype	Physical manifestation of an organism's genotype
Gene	Inherited factor that controls different traits of an organism
Allele	Different versions of a gene that produce different physical traits
Locus	Location of an allele on a chromosome
Heterozygous	Having different alleles for a gene
Homozygous	Having the same alleles for a gene
Epistasis	A gene at one locus affects the expression of a gene at another locus
Incomplete dominance	Alleles are not completely dominant over others resulting in a phenotypic mixing
Pleiotropy	One gene affects phenotype in many ways
Polygenic inheritance	Several genes have an additive affect on one trait; skin color is an example
True-breeding	Self-pollination produces offspring that have the same traits as the parents

CELL CYCLE ABBREVIATED

G0 of interphase	Nondividing period of cell cycle; most human body cells are in this phase
G1 of interphase	Growth period proceeding mitosis; proteins and organelles are produced
S of interphase	Cell growth and duplication of chromosomes; half of cell cycle
G2 of interphase	Chromatin and chromosomes are visible; centrosomes replicated
Prophase	Chrosomes condense; mitotic spindle forms; nucleoli disappear; centrosomes separate
Prometaphase	Microtubules attach to chromatids' kinetochores; nuclear envelop fragments
Metaphase	Chromosomes line up along metaphase plate; longest stage of mitosis
Anaphase	Sister chromatids migrate to opposite poles; shortest stage of mitosis
Telophase	Nuclear envelope reforms around daughter nuclei and less condensed chromosomes
Cytokinesis	Cytoplasm divides; two daughter cells form; cell proceeds through interphase
G1 checkpoint	Regulatory point when cell proceeds to mitosis or G0 phase
G2 checkpoint	Point directly before mitosis; mitosis happens if cyclin accumulates
M checkpoint	Regulatory point during mitosis

MEIOSIS REVISITED

Meiosis I	Homologous chromosomes separate, and two haploid daughter cells form
Meiosis II	Sister chromatids separate, and four haploid daughter cells form
Prophase I	Crossing over; synapsis; spindle forms; homologues move to metaphase plate
Metaphase I	Kinetochores of nonsister chromatids attach to microtubules
Anaphase I	Homologues separate and move to opposite poles
Telophase I	Cytokinesis and formation of two haploid daughter cells; chromosomes are doubled
Prophase II	Spindle forms; sister chromatids move to metaphase plate
Metaphase II	Kinetochores of sister chromatids attach to microtubules
Anaphase II	Sister chromatids separate and move to opposite poles
Telophase II	Cytokinesis and formation of four haploid daughter cells
Synapsis	Homologues line up and are connected by synaptonemal complex
Crossing over	Genetic exchange between nonsister chromatids; form X shaped chiasma
Tetrad	Four nonsister chromatids joined together

CRUNCH KIT
List of Lists

TOP 12 ORGANIC MOLECULES

Glucose	Sugar that functions in energy storage; glycogen and starch are other examples
Cellulose	Structural sugar in plant cell walls; chitin is found in insects' exoskeletons
Fat	Lipid with fatty acid tails that is saturated or unsaturated; hydrophobic
Phosphoslipid	Forms lipid bilayer of plasma membranes; hydrophobic core; hydrophilic exterior
Steroid	Cholesterol is an example found in the plasma membrane; precursor for hormones
Protein	Made of amino acids; functions in immunity, support, and cell signaling
Chaperon	Protein that helps other proteins fold correctly into four structural levels
Enzyme	Protein that speeds up reactions; binds to substrate and lowers activation energy
Receptor	Protein on membrane surface; binding of chemical signal changes its structure
DNA	Nucleic acid with nitrogenous base, deoxyribose sugar, and phosphate group
RNA	Nucleic acid with ribose sugar; single stranded; functions in transcription
ATP	Fuels other reactions and powers cellular processes by hydrolysis

TOP 12 REACTIONS

Dehydration	Macromolecules are made by the removal of water and formation of covalent bonds
Hydrolysis	Breaking of a macromolecule's bonds by adding water
Oxidation	Transfers hydrogen to oxygen to create hydrogen peroxide
Phosphorylation	Transfers phosphate groups between compounds; drives signal transduction
Dephosphorylation	Removes phosphate group from compounds; controls signal transduction
Respiration	Produces ATP to fuel cellular processes; types: substrate-level, oxidative, and photo
Glycolysis	Splitting of glucose into two molecules of pyruvate by removing water
Chemiosmosis	Creates hydrogen gradient across membrane to fuel ATP production
Fermentation	Produces ATP in the absence of oxygen with substrate-level phosphorylation
Photosynthesis	Converts light to chemical energy; consumes water and carbon dioxide
Carbon fixation	Transfer of carbon from inorganic to organic molecule; second state of photosynthesis
Electron transport	Transfer of electrons between compounds; controls energy release and produces ATP

ALL ABOUT RNA

Ribose	Five-pentose sugar found in RNA molecules
Uracil	Replaces thymine as nitrogenous base in RNA molecules
Transcription	RNA synthesis from DNA template
RNA polymerase	Enzyme that unzips DNA and synthesizes RNA transcript
Transcription factors	Proteins that assist RNA polymerase to bind to DNA and carry out transcription
Splicing	Removal of exons and binding of introns to make a functional RNA molecule
Spliceosome	Complex of proteins where RNA splicing takes place
mRNA	Serves as template for protein synthesis
tRNA	Assists mRNA in translation by binding to amino acids
rRNA	Enzyme; structural component of ribosomes
Transcript	RNA molecules before processing
snRNA	Play a structural and enzymatic role in spliceosomes that carry out splicing
SRP RNA	Recognize signal molecules
snoRNA	Assists pre-rRNA processing
siRNA	Involved in gene expression
miRNA	Involved in gene expression

TOP 13 DNA MUTATIONS

Mutagen	Agent that damages DNA by changing structure
Point mutation	Change to a single base pair
Nondisjunction	Chromosome breakage during crossing over or improper separation during meiosis
Base-pair substitution	Replacement of nucleotide and its partner; may alter proteins and affect functioning
Silent mutation	Caused by substitutions; translates the same amino acids; has no affect on proteins
Missense mutation	Caused by substitutions; may not code for correct amino acid
Nonsense mutation	Codon changed to stop codon; creates shorter, nonfunctional protein
Insertion	Nucleotide pair added; alters reading frame; causes missense, nonsense, and termination
Deletion	Nucleotide pair is lost; alters reading frame; causes missense, nonsense, termination
Duplication	Caused by nondisjunction; segment of DNA removed and another segment duplicated
Inversion	Caused by nondisjunction; segment of DNA inverted
Aneuploidy	Nondisjunction causes abnormal number of chromosomes in daughter cells
Translocation	Caused by nondisjunction; segment of DNA moved to nonhomologous chromosome

CRUNCH KIT
List of Lists

TOP 12 PLANT STIMULI

Auxin	Hormone; functions in cell elongation, root growth, and differentiation
Ethylene	Hormone; functions in ripening, leaf loss, and mechanical stress
Brassinosteroid	Hormone; functions in cell differentiation and slows leaf loss
Abscisic acid	Hormone; functions in growth inhibition, closing of stomata, and seed dormancy
Cytokinin	Hormone; functions in division, differentiation, apical dominance, and aging
Gibberellin	Hormone; stimulates germination, cell elongation, and fruit growth
Florigen	Unknown hormone that causes flowering by detecting photoperiod
Phototropism	Plant response to light detected by photoreceptors and phytochromes
Photoperiodism	Response to day and night length
Gravitropism	Plant response to gravity; statoliths cause roots to grow down and stems to grow up
Thigmomorphogenesis	Change in plant structure caused by mechanical stimuli
Thigmotropism	Change in growth in response to touch; allows plant to grow around barrier

TOP 13 ANIMAL ENZYMES

Amylase	Secreted by pancreas and salivary glands; breaks down polysaccharides
Pepsin	Secreted by stomach; breaks down proteins
Trypsin	Secreted by pancreas; breaks down polypeptides
Chemotrypsin	Secreted by pancreas; breaks down polypeptides
Carboxypeptidase	Secreted by pancrease; breaks down small polypeptides
Nuclease	Secreted by pancrease; breaks down nucleic acids
Lipase	Secreted by pancreas; creates fat droplets
Bile salts	Secreted by gall bladder; creates fat globules
Disaccharidase	Secreted by small intestine; breaks down disaccharides
Proteases	Secreted by small intestine; breaks down small peptides
Nucleotidase	Secreted by small intestine; breaks down nucleotides
Nucleosidase	Secreted by small intestine; breaks down nucleosides
Phosphatase	Secreted by small intestine; breaks down nucleosides

TOP 13 PLANT CELLS/TISSUES

Parenchyma cells	Totipotent plants cells that are the least specialized and perform many functions
Sclerids	Sclerenchyma cells that form thick, hard walls
Fibers	Sclerenchyma cells that form slender, strong cells
Collenchyma cells	Structurally flexible with uneven thickness; support young plants
Tracheids	Conduct water; slender, tapered structure; prevent collapse of cells
Vessel elements	Conduct water; wide, thin structure; form continuous vessels
Sieve-tube members	Conduct sugar and other nutrients; have no organelles
Companion cells	Have organelles that support sieve-tube members
Dermal tissue	Cuticle prevents water loss; periderm replaces epidermis in older plants
Vascular tissue	Xylem conducts water and some nutrients; phloem conducts sugars and some nutrients
Ground tissue	Support cells; pith is interior to vascular tissue; cortex is exterior to vascular tissue
Mesophyll	Ground tissue in leaves made up of parenchyma cells that contain chloroplasts
Guard cells	Regulate opening and closing of stomata

TOP 14 ANIMAL CELLS

Fibroblasts	Cells of connective tissue; secrete proteins to create structural network
Neurons	Cells of nervous system that transmit electrical impulses
Glial cells	Nerve cells that support neurons
Muscle cells	Made up of actin and myosin fibers; form skeletal, smooth, and cardiac muscle
Erythrocytes	Red blood cells; carry oxygen and carbon dioxide through the bloodstream
Platelets	Fragments of cells that function in blood clotting; found in the bloodstream
Stem cells	Pluripotent cells in the red bone marrow and other areas; produce other cells
Phagocytes	White blood cells that ingest foreign microbes
Mast cells	Cells of connective tissues; store histamine that functions in inflammatory response
Lymphocytes	Immune cells that recognize specific antigens
Natural killer cells	Immune cells that attack infected and cancerous cells
Neurosecretory cells	Nerve cells that secrete hormones into the bloodstream
Alpha cells	Cells of the pancreas that secrete glucagon
Gametes	Sperm and egg cells; haploid

CRUNCH KIT
List of Lists

TOP 12 PLANT DEFENSES

Avirulent	Pathogen made mildly harmful by plant resistance genes
Elicitors	Molecules that cause a nonspecific response to infection
Gene-for-gene recognition	Plant resistance genes recognize specific pathogens, making them mildly harmful
Hypersensitive response	Localized response based on gene-for-gene recognition that seals off the infected area
Jasmonic acid	Chemical that activates genes important in plant defense
Phytoalexins	Antimicrobial compounds produced as a nonspecific defense to infection
PR proteins	Antimicrobial and warning signal chemicals
Recruitment	Wounding causes the release of signals that attract predators of pathogens
Resistance genes	Plant genes that code for defense proteins that fight a specific pathogen
System acquired resistance	Nonspecific response to infection that protects against many pathogens
Virulent	Pathogen that can kill a plant that doesn't have resistance genes for protection
Warning signals	Infection causes the release of signals that warn neighboring plants

TOP 20 ANIMAL HORMONES

ADH	Increases water retention in kidneys
Oxytocin	Contracts mammary glands and uterus
Growth hormone	Causes growth and increased metabolism
Insulin	Lowers blood glucose levels
Glucagon	Raises blood glucose levels
Androgen	Stimulates male sex characteristics
Estrogen	Stimulates female sex characteristics
Melatonin	Establishes biorhythms
Epinephrine/ Norepinephrine	Increase metabolism; cause fight or flight response
Progesterone	Stimulates growth of the uterine lining
Gastrin	Simulates gastric juice production
Secretin	Stimulates bicarbonate release
Prolactin	Stimulates milk production
Calcitonin	Lowers blood calcium levels
PTH	Parathyroid hormone; raises blood calcium levels
FSH	Stimulates gamete production
LH	Stimulates reproductive organs
Enderogastron	Inhibits peristalsis and HCL secretion
CCK	Stimulates digestive enzyme release
Thyroxine	Maintains metabolism

PLANT GROWTH 101

Meristems	Embryonic tissues that grow throughout a plant's lifetime
Derivatives	Specialized cells produced by the meristems
Initials	Embryonic cells produced by the meristems
Zone of maturation	Area of root growth farthest from the tip; contains differentiated cells and hairs
Zone of elongation	Area of root growth in the middle of the root where cells elongate; pushes root into soil
Zone of division	Area close to the root tip where cells divide; contains meristems and derivatives
Root cap	Protective covering at root tip that secretes lubricant as the root pushes into the ground
Secondary growth	Made by vascular and cork cambium; produces the thick parts of woody plants
Vascular cambium	One cell layer in a cylinder that produces secondary xylem and phloem
Cork cambium	Made by outer cortex and pericycle; secretes suberin; produces cork for protection
Fusiform initials	Form secondary xylem and phloem; oriented parallel to stem and roots
Ray initials	Form parenchyma cells in vascular rays that transport water and nutrients
Secondary xylem	Older xylem tissues; heart wood close to center; sapwood in outer layers
Bark	Secondary phloem and periderm

YOU'VE GOT NERVE

Axons	Transmit electrical impulses; long processes of neuron
Cerebrospinal fluid	Fills central canal of spinal cord and four ventricles of the brain; supplies nutrients
Corpus callosum	Axons that connect brain hemispheres
Dendrites	Receive electrical impulses; branched structures of neurons
Depolarization	An electrical impulse causes the inside of the membrane to become less negative
Hyperpolarization	An electrical impulse causes the inside of the membrane to become more negative
Lateralization	Segregation of different functions in right and left hemispheres of brain
Myelin sheath	Made up of Schwann cells; insulate neurons and help transmit electrical impulses
Neurotransmitters	Chemicals that transmit electrical impulses from presynaptic to postsynaptic cells
Refractory period	Time when action potential can't occur because sodium ion channels are closed
Saltatory conduction	Action potential jumps from one Node of Ranvier to the next
Threshold	Membrane voltage that must be reached to cause an action potential
White/gray matter	Tracts of axons/dendrites and neuronal cell bodies in the brain

CRUNCH KIT
List of Lists

CELLULAR RESPIRATION 101

Reactants	Oxygen and organic molecules such as glucose
Products	Carbon dioxide, water, and energy in the form of ATP
Mitochondria	Site of citric acid cycle and oxidative phosphorylation
Glycolysis	Splitting of glucose produces two molecules of pyruvate; occurs in cytosol
Citric acid cycle	Completes glucose breakdown and produces carbon dioxide
Oxidative phosphorylation	Combines electron transport and chemiosmosis to produce ATP
Substrate-level phosphorylation	Occurs in glycolysis and citric acid cycle; produces ATP
NADPH	Carries electrons from glycolysis and the citric acid cycle to oxidative phosphorylation
Pyruvate	Product of glycolysis; used in citric acid cycle after conversion to acetyl CoA
Acetyl CoA	Formed by the removal of water, oxidation, and catalysis by coenzyme A of pyruvate
ATP synthetase	The enzyme that produces ATP; found in mitochondrial inner membrane
Cytochromo-somes	Proteins that function as electron carriers

PHOTOSYNTHESIS 101

Chloroplasts	Organelles where photosynthesis takes place; surrounded by double membrane
Chlorophyll	Pigment that absorbs light energy and becomes excited, releasing an electron
Light reactions	Splitting of water produces an electron that is transferred to NADP+
Photophosphoryl-ation	Occurs in light reactions; drives ATP production
Photosystems	Harvest light using chlorophyll and proteins bound to thylakoid membrane
Thylakoid membrane	Membrane inside chloroplasts where light reactions take place
Noncyclic electron flow	Electrons travel from photosystem I to photosystem II during light reactions
Cyclic electron flow	Electrons cycle through photosystem I during light reactions
Calvin cycle	Uses ATP produced in light reactions to produce carbohydrates in stroma
Stroma	Fluid-filled space inside chloroplasts
Carbon fixation	Occurs in Calvin cycle; transfers inorganic carbon from carbon dioxide to RuBP
Rubisco	Catalyzes carbon fixation during Calvin cycle; has an affinity for oxygen, as well

TOP 16 BIOMES

Tropical forest	Rain forest near equator with warm temperatures and high biodiversity
Desert	Area of low vegetation, small animals, little rain, and variable temperatures
Savanna	Warm area near equator with scattered trees and large herbivores
Chaparral	Coastal areas with diverse plants, warm summers, and seasonal rains
Temperate grassland	Area with grasses, grazing animals, cold winters, hot summers, and seasonal rainfall
Coniferous forest	Conifers; many birds; droughts; long winters
Broadleaf forest	Closed canopy; many animals; humidity; rain
Tundra	Arctic areas; permafrost; dwarf shrubs; grazing animals; short summers
Lake	Standing body of water
Wetland	Supports aquatic life; many plants
Stream/river	Moving body of water; many fish species
Estuary	Area where river and sea meet; many crustaceans and fish
Intertidal zone	Submerged during high tide; diverse algae
Oceanic pelagic	Open ocean water; currents; phytoplankton
Marine benthic	Below the surface of coastal waters
Coral reef	Tropic marine areas with light; made of calcium carbonate

ECOLOGY IN BRIEF

Population	Species group living in the same area and influenced by the same factors
Community	All living organisms in an area that are close enough to interact with one another
Ecosystem	All biotic and abiotic factors affecting a community
Biome	Large areas influenced by similar abiotic and biotic factors; terrestrial or aquatic
Competition	When two species fight for single resource; one species may be eliminated
Niche	All biotic and abiotic resources a species uses
Resource partitioning	No two species living in the same place can belong to the same niche
Character displacement	Populations in the same geographic area have different traits and resources
Biomass	Dry weight of all organisms belonging to the same species
Gross primary production	All the light energy converted to chemical energy by producers
Carrying capacity	Maximum number of individuals in a population an environment can support
Dominant species	The mass of the total species population that has the greatest biomass
Keystone species	Species that greatly affects an environment

CRUNCH KIT
List of Lists

REMEMBERING DARWIN

Natural selection	Individuals with certain traits outcompete others for resources
Adaptation	The environment changes, and some individuals adapt to the new conditions
Descent with modification	Selection and adaptation alter populations and may create new species
Speciation	Creation of a new species
Homology	Similar traits shared by species with shared ancestry
Analogy	Similar traits shared by unrelated species
Biogeography	Related species are found in the same areas
Fossil record	Fossils of extinct species show how organisms have changed over time
Origin of Species	Darwin's publication that outlined the theory of evolution
Galapagos islands	Islands off the coast of South America visited by Darwin
Beagle	Ship that sailed to South America; Darwin studied plant and animal species there
Artificial selection	Selective breeding of species for certain desired traits
Endemic	Species that are found in only one part of the world

ALL ABOUT EVOLUTION

Gene pool	Genes found in a population
Genetic drift	A finite population size causes allele frequencies to fluctuate
Gene flow	As individuals move in and out of a population, genes are gained and lost
Fitness	Measure of an individual's reproductive success
Relative fitness	Measures a genotype's contribution to the gene pool compared to other genotypes
Sexual selection	Selection of mates affects population's genetic makeup
Neutral variation	Does not increase or decrease fitness
Reproductive isolation	Barriers prevent two species from interbreeding
Allopatric speciation	A population is separated by geographic barriers and forms new species
Sympatric speciation	A population in one geographic area is separated and forms new species
Adaptive radiation	Formation of many new species in a new environment
Punctuated equilibrium	A species changes most just when it evolves from the parent species
Gradualism	Gradual diversion of daughter species

KEYS TO DIVERSITY

Bacteria	Prokaryotes that are classified as gram positive or negative
Archaea	Prokaryotes that may be extremophiles or methanogens
Protists	Unicellular eukaryotes that are abundant and diverse
Bryophytes	Nonvascular plants such as liverworts, hornworts, and mosses
Seedless vascular plants	Lycophytes such as club mosses and pterophytes such as ferns and horsetails
Gymnosperms	Vascular plants with naked seeds
Angiosperms	Vascular plants that produce flowers and seeds enclosed in ovaries
Fungi	Eukaryotes that function in nutrient recycling and decomposition
Invertebrates	Animals that do not have a backbone
Chordates	Animals with a notochord and nerve cord
Craniates	Chordates with a head
Vertebrates	Craniates with a backbone
Tetrapods	Vertebrates with hands and feet
Amniotes	Tetrapods that have an egg surrounded by an amniotic sac
Mammals	Animals that have hair and produce milk
Humans	Hominids that walk upright and have a large brain

CLASSIFICATION 101

Phylogeny	Evolutionary history of a species
Systematics	Classifying organisms with analytical methods
Taxonomy	Using similar characteristics to classify organisms in a hierarchy
Phylogenic tree	Diagram with branches that depict evolutionary relationships
Cladogram	Diagram that depicts the characteristics species share
Clade	An ancestor and its descendents
Cladistics	Grouping species into different clades
Phylogram	Diagram with branches that reflect changes in DNA sequences over time
Ultrametric tree	Phylogram that is related to the fossil record
Parsimony	Classification of species that requires the fewest evolutionary events
Molecular systematics	Classification based on similarities in DNA or RNA sequences
Homologous genes	Genes from different species with shared ancestry
Molecular clock	Measures evolution by assuming that some DNA changes at a constant rate
Neutral theory	Most changes in DNA and proteins does not affect fitness
Binomial	Scientific name for species

FINAL TIPS AND ABOUT THE AUTHOR

FINAL TIPS

- Focus your energy on the commonly tested areas (animals, plants, evolution, genetics, cells)
- Don't neglect or spend all of your time on other topics (diversity, phylogeny, chemistry)
- Memorizing terms isn't as important as understanding concepts
- Remember the most important themes in biology
- Big tests are no fun no matter how hard you study
- Remember that you've already learned these topics in class and studying is a refresher
- Don't neglect global issues such as global climate change and loss of biodiversity
- Go over the pie chart at the beginning of the Cram kit to see which subjects are most important
- Biology is the study of life, so don't neglect yourself while you're studying!

ABOUT THE AUTHOR

In high school, after taking every biology class that she possibly could, **Katherine Lutz** went on to college to study—surprisingly enough—biology. Since graduating from **Wellesley College** in **1998**, Katherine has practiced, taught, and written about science. After several years working in a lab, she received a graduate degree in science journalism. Jobs at NPR and the *Boston Globe* followed, but her first love is teaching. She developed educational materials for tech savvy AP Biology teachers but isn't much of a techie herself. In fact, she only just started texting—what does lol mean anyway? According to her students, she's the only technologically challenged, sci-fi fanatic geek.

ABOUT THE EDITOR

DEAN SCHAFFER

Since leading Los Angeles's Taft High School to a national Academic Decathlon championship and shipping off to Stanford University, Dean Schaffer has designed and developed DemiDec's signature Power Guides and Cram Kits. Over the years, he has maintained his affinity for a variety of metaphorical hats (editing, writing, and layout, especially), non-metaphorical sunglasses (Aviators, always), and the guitar (Fender, usually). When he's not editing, Dean is generally looking for something nearby to edit—dont we all luv to find speling erors?

In his spare time, Dean ponders whether he'll ever be able to handle the luxury of spare time; luckily, he avoids this metaphysical quandary altogether by choosing not to affiliate himself with relaxation of any form. Instead, he occupies himself with songwriting, playing guitar, and parallel structure-ing. When he isn't doing those things, he's considering the merits of democratic elections, oligarchic disinterestedness, and delicious gouda cheese.

ABOUT DEMIDEC

THE DEMIDEC STORY

Since 1994, DemiDec has been the worldwide leader in student-centered learning experiences, from its curriculum for the Academic Decathlon in the United States to its college prep academies in Asia. DemiDec now brings its unique approach—and its mascot, the alpaca—to a whole new realm: AP and SAT materials.

To learn more about DemiDec and how you can help Spread the Cram, visit www.demidec.com.

ABOUT THE WORLD SCHOLAR'S CUP

Six subjects. Four events. Teams of three.

DemiDec's World Scholar's Cup is an international team academic tournament with thousands of participants in 30 countries. It centers on a different them each year—from the Frontier to the Fall of Empires. Events include team debate, a scholar's bowl, and more.

Discover this year's theme, learn more about the program, and sign up for free at www.scholarscup.org.

Made in the USA
Monee, IL
21 April 2023

32214700R00067